Software Engineering Reviews and Audits

BOOKS ON SOFTWARE AND SYSTEMS DEVELOPMENT AND ENGINEERING FROM AUERBACH PUBLICATIONS AND CRC PRESS

Design and Safety Assessment of Critical Systems
Marco Bozzano and
Adolfo Villafiorita
978-1-4398-0331-8

Implementing and Developing Cloud Computing Applications
David E. Y. Sarna
978-1-4398-3082-6

Secure Java: For Web Application Development
Abhay Bhargav and B. V. Kumar
978-1-4398-2351-4

Scrum Project Management
Kim H. Pries and Jon M. Quigley
978-1-4398-2515-0

Engineering Mega-Systems: The Challenge of Systems Engineering in the Information Age
Renee Stevens
978-1-4200-7666-0

Certified Function Point Specialist Examination Guide
David Garmus, Janet Russac, and
Royce Edwards
978-1-4200-7637-0

Enterprise Systems Engineering: Advances in the Theory and Practice
George Rebovich, Jr.
and Brian E. White
978-1-4200-7329-4

Process-Centric Architecture for Enterprise Software Systems
Parameswaran Seshan
978-1-4398-1628-8

Secure and Resilient Software Development
Mark S. Merkow and
Lakshmikanth Raghavan
978-1-4398-2696-6

Real Life Applications of Soft Computing
Anupam Shukla, Ritu Tiwari,
and Rahul Kala
978-1-4398-2287-6

Product Release Planning: Methods, Tools and Applications
Guenther Ruhe
978-0-84932620-2

Process Improvement and CMMI® for Systems and Software
Ron S. Kenett and Emanuel Baker
978-14200-6050-8

Applied Software Product Line Engineering
Kyo C. Kang, Vijayan Sugumaran,
and Sooyong Park
978-1-42006841-2

CAD and GIS Integration
Hassan A. Karimi and Burcu Akinci
978-1-4200-6805-4

Applied Software Product-Line Engineering
Kyo C. Kang, Vijayan Sugumaran,
and Sooyong Park, eds.
978-1-4200-6841-2

Enterprise-Scale Agile Software Development
James Schiel
978-1-4398-0321-9

Handbook of Enterprise Integration
Mostafa Hashem Sherif, ed.
978-1-4200-7821-3

Architecture and Principles of Systems Engineering
Charles Dickerson, Dimitri N. Mavris,
Paul R. Garvey, and Brian E. White
978-1-4200-7253-2

Theory of Science and Technology Transfer and Applications
Sifeng Liu, Zhigeng Fang,
Hongxing Shi, and Benhai Guo
978-1-4200-8741-3

The SIM Guide to Enterprise Architecture
Leon Kappelman
978-1-4398-1113-9

Getting Design Right: A Systems Approach
Peter L. Jackson
978-1-4398-1115-3

Software Testing as a Service
Ashfaque Ahmed
978-1-4200-9956-0

Grey Game Theory and Its Applications in Economic Decision-Making
Zhigeng Fang, Sifeng Liu,
Hongxing Shi, and Yi LinYi Lin
978-1-4200-8739-0

Quality Assurance of Agent-Based and Self-Managed Systems
Reiner Dumke, Steffen Mencke,
and Cornelius Wille
978-1-4398-1266-2

Modeling Software Behavior: A Craftsman's Approach
Paul C. Jorgensen
978-1-4200-8075-9

Design and Implementation of Data Mining Tools
Bhavani Thuraisingham, Latifur Khan,
Mamoun Awad, and Lei Wang
978-1-4200-4590-1

Model-Oriented Systems Engineering Science: A Unifying Framework for Traditional and Complex Systems
Duane W. Hybertson
978-1-4200-7251-8

Requirements Engineering for Software and Systems
Phillip A. Laplante
978-1-4200-6467-4

Software Engineering Reviews and Audits

Boyd L. Summers

CRC Press
Taylor & Francis Group
Boca Raton   London   New York

CRC Press is an imprint of the
Taylor & Francis Group, an **Informa** business

AN AUERBACH BOOK

Auerbach Publications
Taylor & Francis Group
6000 Broken Sound Parkway NW, Suite 300
Boca Raton, FL 33487-2742

© 2011 by Taylor and Francis Group, LLC
Auerbach Publications is an imprint of Taylor & Francis Group, an Informa business

No claim to original U.S. Government works

Printed in the United States of America on acid-free paper
10 9 8 7 6 5 4 3 2 1

International Standard Book Number: 978-1-4398-5145-6 (Hardback)

Visit the Taylor & Francis Web site at
http://www.taylorandfrancis.com

and the Auerbach Web site at
http://www.auerbach-publications.com

Contents

List of Figures

List of Tables

Preface

I have worked in systems engineering, software development, software configuration management, software testing, and software quality assurance organizations for more than 30 years. My experience in these selected fields has been somewhat different from that of most people. I always wanted to experience the software disciplines required in each area of expertise. I know that many individuals or groups reading this book will be surprised to know that successful software engineering reviews and audits are beneficial to the success of software industries and military and aerospace programs. The commercial software world can benefit from this book by helping companies realize that they can succeed in this crazy and sometimes confusing software market and by being aware that effective reviews and audits for software will help them to be more successful.

I attended college and majored in business management with an emphasis in information systems, production and operations management, quantitative analysis methods, statistical analysis, computer science, and application programming. During this time, I worked for an aerospace company and began my journey into software development as a senior systems analyst. I dove into software requirements, software design, code and unit testing, configuration control, software builds, software deliveries, and providing software documentation supporting each area of expertise. I have been involved in performing numerous software engineering reviews and audits to ensure compliance with contractual requirements.

In leading multiple software engineering development teams, I continually tackled complex technical challenges to ensure that system/software engineering problems were addressed and resolved. The main objective of the technical leadership role for software design teams, through the use of common software tools for both UNIX® and Windows® platforms, was to ensure that common software tools were institutionalized. In order to establish and implement common software tools, the necessary capabilities must be provided for teams to learn from other software groups using similar software tools and processes. Adapting these new and proven tools will decrease the software build flow times from days to hours, so teams can be more responsive to customer needs, issues, and concerns. Software metrics should be provided weekly and monthly to senior management for problems and software build time evaluations to ensure that issues are addressed and resolved. The software tasks for engineering reviews and audits performed during the development lifecycle prior to deliveries and formal audits will improve quality and execution.

Aerospace programs present challenges for personnel working in key technical roles. An example is a replacement of key multiplexer computer systems for the United States Air Force (USAF). Technical software issues are addressed and coordinated daily for the customer. Software teams are directed to capture software baselines and ensure that engineering reviews and audits for software have been performed for installations of new software baselines during test and integration activities.

My career culminated with moving to numerous military and aerospace programs and being involved in the Capability Maturity Model Integration® (CMMI®) activities for systems engineering and software development. I supported and participated in CMMI audit teams and performed CMMI appraisals to certify military and aerospace software programs for accomplishing Level 2 and Level 5 ratings for capability certifications.

SUMMARY

Before software, military, and aerospace programs implement software engineering reviews and audits, it is important to understand the software development lifecycle. Chapters are included that define methods for systems engineering, software design, software quality assurance, software configuration management, and software suppliers/subcontractors. The intent of this book is to ensure that software engineering reviews and audits are conducted and performed incrementally in software development schedules. The implementation of review and audit disciplines will benefit software companies and military and aerospace programs and ensure that formal audits (i.e., FAI, FCA, and PCA) are successful the first time.

Acknowledgments

In the last 30 years of working and gaining experience in software industries and military and aerospace programs, I have been motivated to write a book related to the understanding the importance of performing effective software engineering reviews and audits. My career as a software engineer has included designing, controlling, building, installing, and validating software, which is my passion. Before a company for which I work delivers software to its customers, both in the United States and internationally, I feel a sense of accomplishment that my software has successfully passed critical software reviews and major formal audits. The experience I gained working in multiple data centers for commercial banking, Hercules Aerospace, Mini-System software contractors, and the Boeing Company for 30 years has given me the knowledge of numerous software process improvement activities. It is outstanding to work with the software teams with which I am associated. Senior management and software managers have always been a support to me, and I greatly appreciate their guidance. This book has been on my mind for 30 years, due to seeing the good and the bad of performing effective software engineering reviews and audits.

My acknowledgments go to all the second-level software managers I've worked with for allowing me to excel in the commercial software world and military and aerospace programs. In my early years, and with the support of my wife and family, I received the Outstanding Young Man of America Award and achieved a Bachelor of Science (BS) in business management with an emphasis on computer science while attending Weber State University. Thank you to my lovely wife, for her support and patience with me throughout these past 34 years.

After college the Hercules Aerospace Division employed me as a senior software designer. The software team I worked with gave me insight to the software development environment. I will never forget when my first software manager told me to go work for Boeing and sharpen my software skills. I was soon employed by the Boeing Company in the Software Configuration Management (SCM) organization supporting the B2 and F-22 Raptor programs. The software manager and leads were instrumental in my areas of expertise for advancement in the Boeing Company. Special recognition goes to the Boeing software managers in Denver, Colorado, for allowing me the opportunity to expand my skills in CMMI, software design, software configuration management, and software quality assurance methods. The F-22 Raptor and Airborne Early Warning & Control (AEW&C) software managers are an inspiration and have given me the opportunity to organize and establish software configuration management teams. Currently I am a software engineer for supporting quality assurance activities and provide software expertise to the F-22 Raptor Air Vehicle System (AVS) and AEW&C programs.

Examples or outlines are made available in the appendices to show how software plans could be defined and documented. Just think about this! Important software lifecycle disciplines are defined and discussed all through this book.

Let's get started!

Boyd Lambert Summers
Maple Valley, Washington

1

Introduction

Software engineering reviews and audits provide the framework and detailed requirements for verifying/validating development efforts. Performing reviews and audits that are successful will ensure achievement in all specified requirements for software design, testing, configuration control, and quality to released configuration baselines. The reviews and audits will improve the individual and team efforts in maintaining a professional setting where software is developed for profit, cost reduction, and service quality improvement.

There is no clear-cut approach to performing software engineering reviews and audits for multiple software companies and military and aerospace programs. Ideas, suggestions, standards, and concepts are adopted or implemented to improve the quality of software management, software development efforts, supplier deliveries, and customer expectations. It is frustrating and confusing when it is time to perform software engineering reviews and audits. We always ask ourselves, what is required, who needs to participate, and what results do we receive for performing these engineering reviews and audits?

I know many of you reading this book will say to yourself, why do software programs need to have software engineering reviews and audits performed? It is for the assurance that software products have been reviewed, audited, and verified, and that they meet required quality requirements. The software designers/developers shake in their boots when it is time for a review and audit of their software. Here comes software quality, and here we go again. The required quality requirements are accomplished so the working world will utilize the delivered software products in the private sector or programs. I sound like a news reporter from Fox News.

Performing quality reviews and configuration management audits will provide the necessary experience for auditors or appraisers to perform software engineering reviews and audits. These software disciplines play key roles in performing effective software engineering reviews and audits. The elements that should be understood by the reader are defined plans for systems engineering, software development, software configuration management practices, and software quality evaluation.

1.1 SENIOR MANAGEMENT

When it is time to perform required or contractual audits (i.e., internal, formal, etc.), it is necessary for senior management to have in place an allocated budget, committed schedules, and trained personnel to perform the required software engineering reviews and audits. There are times when ideas, suggestions, standards, and concepts are adopted to improve software management performance, development efforts, supplier deliveries, customer expectations, and the quality of those efforts. When it comes time to perform software engineering reviews and audits, the following questions are asked:

- What are the requirements?
- Who needs to participate?
- What are the expected results?

This is a start to performing a software engineering review or audit; you need senior management support. The senior management teams must be out there championing the cause. Because teams represent a major shift in the way an organization produces software products and values its people, buy-in and support of senior management is critical. As software teams become more capable and confident, senior managers can delegate the "how-to" to others in the organization, offering much-needed support and encouragement.

Software development and the disciplines required are a dream, but software engineering reviews and audits are serious.

—**Boyd L. Summers**

1.2 PROCESS AND QUALITY

Many software industries and military and aerospace programs have found issues and concerns by effectively performing software engineering reviews and audits. The reviews and audits for software are beneficial when it is time to prepare for informal and formal audits during critical program or project schedules.

It is important to provide detailed information and the steps required to perform effective, constructive, and compliant reviews and audits. Artifacts (i.e., plans, processes, procedures, test reports, data, etc.) are required for these specific reviews and audits. Software companies and military and aerospace programs want to know the requirements for performing software engineering reviews and audits. These reviews and audits ensure that informal and formal audits are ready to be conducted and performed properly and in compliance with software industry and military and aerospace program requirement standards. The important factor for reviews and audits to be successful is having a combined effort of a process model and quality management system working together.

As depicted in Figure 1.1, Relationships of Process and Quality, ensuring compliance to software requirements, software engineering reviews, and software audit processes is based on a process model and a quality management system understanding.

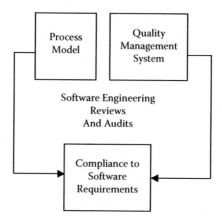

FIGURE 1.1
Relationships of process and quality.

1.3 SOFTWARE LIFECYCLE

The benefit of a software lifecycle is significant due to conducting and preparing for successful software engineering reviews and audits. Software industries and military and aerospace programs improve productivity and meet the demands of customers by having an understanding of a software lifecycle. The roles and responsibilities of software disciplines during the software lifecycle will provide adequate preparation for software engineering reviews and audits. In developing software there are concerns with process and quality. Chapters 2 and 3 will define the process model, quality management system, and the importance of these disciplines in supporting software engineering reviews and audits.

The software lifecycle is a period of time that starts when a software product is conceived and ends when the software product is no longer available for use. The software lifecycle includes a requirements phase, design phase, implementation phase, test phase, installation and checkout phase, and operation and maintenance phase.

1.3.1 Requirements Phase

The requirements phase is the process of defining the software architectures, components, modules, interfaces, and data for a system to satisfy specified system requirements. System documentation conveys requirements, design philosophy, design details, capabilities, limitations, and other characteristics to show reliability. System engineering testing is the process of verifying that specified requirements are in compliance.

1.3.2 Design Phase

The design phase begins with the decision to develop a software product and ends when the product is delivered. This phase involves user needs being translated into software requirements that are transformed into design, implementation of code, testing of code, documentation, and certification for operational use. The software development plan will provide a project plan for the development of the software products.

1.3.3 Implementation Phase

The implementation phase is the period of time in the software lifecycle during which a software product is created from design documentation and debugged. Implementation includes any requirement that impacts or constrains the implementation of a software design, for example, design descriptions, software development standards, program language requirements, software configuration management standards, and software quality standards. Independent verification and validation of software products is performed by software quality assurance, which is independent from the software design organization. A software quality assurance plan is developed to ensure that software quality is applied during the software lifecycle. During the implementation phase software configuration management identifies and defines configuration items, control releases, software changes, records, report change processing, and implementation status throughout the software lifecycle. Functional and physical characteristics are identified and documented. A software configuration management plan is documented to ensure that software configuration control is implemented during the software lifecycle.

1.3.4 Test Phase

The test phase is a period of time in the software lifecycle during which components of a software product are evaluated and integrated based on whether or not software requirements have been satisfied. Test data and associated test procedures exercise program paths to verify compliance to specified software requirements. The test plan describes the approach intended in the test phase. The plan identifies the items to be tested, the testing performed, test schedules, reporting requirements, evaluation criteria, and risk management. Test procedures provide detailed instructions for setup, operation, and evaluation results.

1.3.5 Installation and Checkout Phase

The installation and checkout phase in the software lifecycle is the period of time during which a software product is integrated into an operational environment and tested in the environment to ensure that the software performs as required.

1.3.6 Operation and Maintenance Phase

In the software lifecycle, the operation and maintenance phase will ensure that the software product is employed in the operational environment, monitored for performance, and modified as necessary to correct software problems or to changing requirements. The operational reliability of a software subsystem may differ from reliability in specified test environments.

1.3.7 Conclusion

The software lifecycle phase is the framework and provides an understanding of software engineering processes. The relationship between the process model and quality management ensures that software engineering is compliant and meets requirements. Performing effective software engineering reviews and audits during the phases of the software lifecycle will benefit future preparation towards informal and formal audits.

FURTHER READING

IEEE STD 028-1988 IEEE Standard for Software Reviews and Audits.
IEEE STD 1074.1-1995 IEEE Guide for Developing Software Life-Cycle Processes.
Military Standard (MIL-STD) 1521b (USAF)—Technical Reviews and Audits for Systems, Equipments, and Computer Software, June 4, 1995.

2

Process Model

This chapter describes the components of the process model and each process area. The process areas, when implemented on programs or projects, satisfy goals that are considered to be important to and consistent with the software development lifecycle activities. The process model supports the performance of software engineering reviews and audit activities. Many technology institutes for software have told me that process models have nothing to do with performing these reviews and audits. I beg to differ. Personally, the process model is beneficial for software auditors to utilize when it is time to conduct and perform scheduled software engineering reviews and audit scheduled activities.

2.1 PROCESS MODEL

I had the distinct privilege of meeting Watts S. Humphrey in the early 1990s at a Capability Maturity Model® (CMM®) conference in Seattle, Washington. At that time, I was working for the Boeing Defense and Space Company in Denver, Colorado. I will never forget Mr. Humphrey's profound statement: "The ability of engineering teams to develop quality software products efficiently and effectively greatly depends on the individual engineer." Adopting the Capability Maturity Model Integration (CMMI) will institute principals of the process model for improvement and implement the following:

- An improvement effort is focused on achieving goals and helping software programs become successful.

- Results will be completed within a meaningful time frame with senior management being involved and working with software process owners.
- Significant values and improvements are accomplished in weeks and months instead of waiting for yearly results.

My career culminated with moving to numerous military and aerospace programs and being involved in CMMI activities. Software engineering reviews and audits were performed on a continual basis during critical delivery schedules. By adopting CMMI and performing reviews and audits, software build flow times improved from days to hours. The software development teams were able to have more time to perform (i.e., design, test, integration, etc.) and exceed critical program schedules. I supported and participated in audit teams and performed CMMI appraisals to certify military and aerospace software programs for accomplishing Level 2 and Level 5 ratings for capability certifications.

2.2 CMMI® UNDERSTANDING

The reason I enjoy working in this process model is the understanding of integration processes for the systems and the software engineering environment. It is the continual emphasis on "repeatable processes." I enjoy implementing the process area "Verification." The peer review process defined in Verification supports me when it is time to audit the processes. I work with software design engineers, discussing with many software review boards the software problems and fixes that support multiple military software programs. I am considered the quality enforcer that provides software quality to ensure that the process, such as peer reviews, is followed.

It is a requirement for the software design engineers to conduct an informal or formal peer review in the software review boards. If a formal peer review is conducted, the peer review identification is referenced in a software problem. If an informal peer review is conducted, there is no requirement to enter the peer review identification. You would be amazed at how many software problems are informal. In the software review board, software designers look at me as quality control to ensure that the peer review process is followed each time a software problem is reviewed. The software problem can be moved for more analysis, approved, implemented, rejected, or closed. The peer review is the review of software work products developed by other software designers during the development of those work products in order to identify and fix defects. The peer review is an important and effective verification method implemented via inspections, structured walkthroughs, or a number of other software review methods.

Software designers know how to work with requirements and write effective code, and they take pride in being considered excellent or exceptional at their profession. The peer review process is a repeatable process that ensures that other software designers, along with a software quality person, will review the code to ensure its compliance with coding standards. High software standards enforce the implementation of software processes to ensure that defects or issues are resolved early in software development stages. Process models provide a focus on sound software practices and artifacts to ensure excellence.

The improvements year after year in understanding the model include very good knowledge of its maturity and clarity to software processes. At certain points in a software development lifecycle, improvements are important to ensure that defined software processes are followed. In appraisals, I have noticed that when it is time for a scheduled software engineering review and audit for software, the criteria for these reviews and audits are understood and successful. The reason for these reviews and audits for software is to show that processes are integrated in the minds of others, as well as the ability to show results when answering audit questions, reviewing required documentation, and showing appraisers' or auditors' evidence that software processes are compliant.

2.3 PROCESS AREAS

As a software auditor, I use staged software processes when conducting or performing software engineering reviews and audits. The process areas reflect maturity levels for Managed, Defined, Quantitatively Managed, and Optimizing. They are:

Level 2: Managed
- Requirements Management (REQM) – Establishes responsibilities and activities for management of software requirements.

- Project Planning (PP) – Defines required project planning activities.
- Project Monitoring and Control (PMC) – Defines procedures for reporting the progress of project.
- Supplier Agreement Management (SAM) – Defines selecting qualified software subcontractors, managing them effectively, and tracking performance and results.
- Measurement and Analysis (MA) – Develop and sustain a measurement capability that is used to support management needs.
- Process and Product Quality Assurance (PPQA) – Defines procedures for ensuring that programs or projects meet quality criteria. *Comment*: The first question from my current software manager to me is always, has the program, project, or supplier developed and released a Software Quality Assurance Plan (SQAP) in order to perform engineering reviews and audits?
- Configuration Management (CM) – Defines the procedures for establishing and maintaining software control.

Level 3: Defined

- Requirements Development – Produce and analyze customer, product, and product component requirements.
- Technical Solution – Design, develop, and implement solutions to requirements.
- Product Integration – Assemble the product from the product components; ensure that the product, as integrated, functions properly; and deliver the product.
- Verification – Ensure that selected work products meet their specified requirements.
- Validation – Demonstrate that a product or product component fulfills its intended use when placed in its intended environment.
- Organizational Process Focus – Establish the organizational responsibility for software process activities that improve the organization's overall software process capability,
- Quantitative Process Definition + Integrated Product and Process Development (IPPD) – Develop and maintain a usable set of software process assets that improve process performance across the

projects and provide a basis for cumulative, long-term benefits to the organization.

- Organizational Training – Develop the skills and knowledge of people so they can perform their roles effectively and efficiently.
- Integrated Project Management +IPPD – Establish and manage the project and involvement of the relevant stakeholders according to an integrated and defined process that is tailored from the organization's set of standard processes.
- Risk Management – Identify potential problems before they occur so that risk-handling activities can be planned and invoked as needed across the life of the product or project to mitigate adverse impacts on achieving objectives.
- Decision Analysis and Resolution – Analyze possible decisions using a formal evaluation process that evaluates identified alternatives against established criteria.

Level 4: Quantitatively Managed

- Organizational Process Performance – Establish and maintain a quantitative understanding of the performance of the organization's set of standard processes in support of quality and process-performance objectives, and provide the process-performance data, baselines, and models to quantitatively manage the organization's projects.
- Quantitative Project Management – Quantitatively manage the project's defined process to achieve the project's established quality and process-performance objectives.

Level 5: Optimizing

- Organizational Innovation and Deployment – Select and deploy incremental and innovative improvements that measurably improve the organization's processes and technologies.
- Casual Analysis and Resolution – Identify causes of defects and other problems and take action to prevent them from occurring in the future. *Comment*: The purpose of this book is to identify problems and resolve them to ensure that future formal audits run smoothly, before actual delivery to the customer. To help prevent these problems, conducting and performing software engineering reviews and audits incrementally in the program or project's schedule will eliminate those issues and concerns. Trust me!

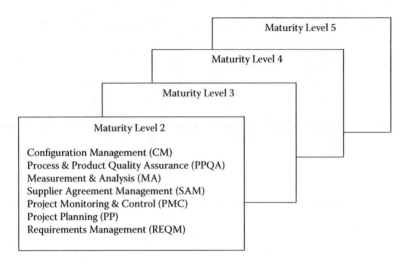

FIGURE 2.1
Process areas in a staged representation.

2.4 STAGED MATURITY LEVELS

The staged maturity level encourages teams to look at process areas in the context of the maturity level of which they are part. The process areas are organized by maturity levels to ensure that this concept is institutionalized in companies, programs, or projects. There is a predetermined path of improvement from Level 2 to Level 5 to achieve the goals of each process improvement. Figure 2.1 shows how process areas are used in a staged representation.

2.5 COMMON FEATURES

Each of the key process areas is organized according to common features. These common features are attributes that indicate whether the implementation and institutionalization of the key process areas are effective, repeatable, and lasting. The five common features are listed below:

1. Commitment to Perform – Describes the actions the organization must take to ensure that the process is established and will endure. Includes practices on policy and leadership.

2. Ability to Perform – Describes the preconditions that must exist in the project or organization to implement the software processes competently. Includes practices on resources, organizational structure, training, and tools.
3. Activities Performed – Describes the roles and procedures necessary to implement a key process area. Includes practices on plans, procedures, work performed, tracking, and corrective action.
4. Measurement and Analysis – Describes the need to measure the process and analyze the measurements. Includes examples of measurements.
5. Verifying Implementation – Describes the steps to ensure that the activities are performed in compliance with the process that has been established. Includes practices on management reviews and audits.

2.5.1 Software Engineering Process Group

The Software Engineering Process Group (SEPG) is assigned responsibility for software process activities. It will establish, manage, and improve the software development processes and the procedures that document the processes in support of Key Process Areas (KPAs) pertaining to Software Engineering Institute (SEI) Capability Maturity Model Integration (CMMI). The objective is to meet the requirements of the SEI CMMI for the reduction of software development risks, schedule, and costs while improving quality and customer satisfaction.

The SEPG operates with appropriate functional area personnel and software personnel from each program. It serves as the software process improvement center for software engineering. It has a chairperson, focal points, SCM members, SQA members, and software program members. The program or project SEPGs operate within each program with program software personnel. They serve as the software process improvement foundation for the software program. The SEPG performs the following functions:

- Process Definition
  1. Develop software processes and procedures for the entire software lifecycle that comply with IDS software engineering standards and procedures, comply with contractual requirements, support ISO 9001 and AS9100A requirements, and address the capability criteria of SEI/CMMI for software development.

2. Evaluate company and program best practices to promote best practices in software engineering processes and procedures.
3. Obtain software engineering functional manager approval of software engineering command media for release.
4. Develop additions and improvements to the software engineering development processes and procedures that will improve software quality and customer satisfaction and reduce the development risks, schedule, and cost.
5. Establish and use a process for receiving, evaluating, and acting upon and reviewing results for proposed processes, procedures, and technology changes.
6. Obtain software engineering functional manager approval of software engineering procedure changes for release into the policy and procedure system.

2.5.2 Process Change Management

Process Change Management begins with a request for a change to an existing process or tool, or a proposal for a new process or tool. The request may be initiated by an Organizational Audit staff member or a Strategic Initiative Team. This process details the steps required to obtain approval for the proposed change.

This process may be triggered in several ways:

- When an SEPG has received input from a software team to implement its process improvement
- When an audit with a process improvement suggestion submits its ideas to be reviewed and worked
- When an employee or manager requests a process change

The requestor submits a request to an organization to formally present the proposed process change to an SEPG. The request should include appropriate justification or substantiation per the required guidelines, and the information should be submitted in a format suitable for presentation. If there is a cost associated with the process change, a cost impact should also be included. If more information is required, the SEPG will contact the originator of the request. Upon hearing the details of the proposed process change, the SEPG will assess the impact and determine if and when the change should be implemented. If there is a cost associated with the change that exceeds the budget plan, then the SEPG will defer the decision to management. If management approves implementation, then a date of completion and implementation will be updated during the next software lifecycle.

FURTHER READING

Ahern, D.M., Clouse, A., and Turner, R. *CMMI Distilled, A Practical Introduction to Integrated Process Improvement,* , Boston, MA: Addison-Wesley, Pearson Education, Rights and Contracts Department, 2001.

Chriss, M.B., Konrad, M., and Shrum, S. *CMMI, Version 1.2, Guidelines for Process Integration and Product Improvement, Peer Review Defects and Removal, Second Edition,* Boston, MA: , Addison-Wesley, Pearson Education, Rights and Contracts Department, 2007.

Curtis, B., Hefley, W.E., and Miller, S.A., *The People Capability Maturity Model, Guidelines for Improving the Workforce,* Boston, MA: Addison-Wesley, Pearson Education, 2002.

Humphrey, W. S. , *Capability Maturity Model (CMM),* Reading, MA: Addison-Wesley Longman, 1995.

Humphrey, W. S., *Introduction to the Team Software Process. TSP:SM,* Reading, MA: Addison-Wesley Longman, 2000.

3

Quality Management Systems

The standards AS9100, SAE AS9110, and ISO 9001 for Quality Systems make up the model for quality assurance in requirements, design, development, production, installation, and service. This model is used when the supplier must ensure conformance to requirements during several stages, which include design, development, production, installation, and service.

3.1 QUALITY MANAGEMENT SYSTEMS

To have quality management in place means simply having documented software processes, executing those processes with knowledgeable employees, monitoring or measuring those software processes, and making continual improvements. The improvements are as follows:

- Plan – Processes are documented to deliver results.
- Do – Implementation is accomplished by a skilled work force.
- Check – Check compliance to improve performance.
- Act – Take actions to continually improve performance.

In order to implement quality management, the customer must be focused, process based, and improvement oriented. Say what you do, do what you say, prove it, and improve it.

Quality management is deployed in software industries and military and aerospace programs to establish requirements and responsibilities for implementation. The standards for military and aerospace programs to adopt and deploy include the following:

- Business units, programs, locations producing air or space vehicles, or weapon systems, are required to comply with AS9100.
- Business units, programs, locations providing maintenance, repair, overhaul, modification services for air and space vehicles or weapon systems are required to comply with AS9100 or may select SAE AS9110.
- Software programs whose products are to be required comply, at a minimum, with ISO 9001.
- Compliance to requirements for the application of AS9100 and ISO9001 on software programs:
  - AS9100 – Quality management system requirement specifies additional requirements for supporting quality management systems for the software industries.
  - SAE AS9110 – Requirements established for software programs will implement a Quality Maintenance System.
  - ISO 9001 – Requirements for a quality management system that can be used for internal application certifications and contractual purposes. It focuses on the effectiveness of the quality management system in meeting customer requirements.

3.2 QUALITY MANAGEMENT SYSTEMS UNDERSTANDING

A general requirement for software quality management is to understand software industries and software programs that establish, document, implement, and maintain effective quality management and will continually improve their effectiveness. These industries and software programs identify the software processes needed for effective quality management and their applications throughout organizations. Quality management concepts determine the sequence and interaction of software processes along with criteria methods to ensure that operation and control of these processes are effective. To ensure the availability of resources and information necessary to support the operation and monitoring of these software processes, measurements and the analysis of these processes are necessary for quality management.

Senior Management provides evidence of its commitment to software development and implementation of software quality management improvement by

- Communicating the importance of meeting customer requirements
- Establishing an effective software quality policy
- Ensuring that software quality objectives are established
- Ensuring that software engineering reviews and audits are conducted and performed

3.3 AS9100 PROGRAM AUDIT

The level of involvement and commitment of senior management in relation to overall quality management in supporting AS9100 program audits is as follows:

- Preparation and implementation of the internal audit processes
- Communication with the organization being audited
- Oversight of specific quality management requirements throughout the organization

3.4 FUTURE QUALITY MANAGEMENT SYSTEMS OBJECTIVES

I like to reference ISO 9001 disciplines and requirements during software engineering review and audit activities. Before implementing these reviews and audits for software, it is important to have future software quality management objectives in place. To ensure these objectives for software, senior management will provide:

- That the planning for quality is carried out in order to meet software requirements
- That the integrity of software quality is maintained when changes are planned and implemented
- The organization to achieve and sustain quality of the product or service produced so as to meet the customer's stated or implied needs
- An organization to provide confidence to management that the intended quality is being achieved and sustained

- Confidence to the customer that quality is being achieved in the delivered product provided

---

3.5 QUANTITATIVE MANAGEMENT PROCESS

All organizations, programs, and projects perform quantitative management using a standard Quantitative Management Process (QMP) along with work instructions. Collection and analysis of standard and specific metrics are usually described in a QMP. Software metrics will be used on the program to manage the software development activity. These metrics are used to evaluate the maturity of the software, measure progress of the development and test efforts, and identify software risks.

FURTHER READING

AS9100 Aero Space (AS) Standard Quality Management System Requirements Guidelines for the Application—Part 2 – 1997.

SAE AS standard, Quality Maintenance System Requirements for Aviation, Space, and Defense Organizations—2009-01.

4

Policy

A policy is the key element in the engineering process, and there are organizational, software planning, and control procedures to support these key elements. The significant activities are defined in this chapter. To conduct a successful software development program, one should understand the scope of the work to be accomplished.

4.1 POLICY UNDERSTANDING

A policy provides a mission statement of direction and guidance for software industries and military and aerospace programs. Policies are the highest level of authority and are consistent with the vision one should use to be successful.

In the past I have seen in many software companies and military and aerospace programs that policies are not reviewed, implemented, or even considered in the work place. Why? I stated in the first paragraph that a policy is a mission statement and should be used as a starting point for performing software development and supporting software engineering reviews and audits. A very effective policy I like to review over and over is a policy for software quality. The software quality policy states that we are the difference, such as

- I am personally responsible and accountable for the quality of my work.
- I acquire/use the necessary tools and skills needed to meet quality requirements.
- I know my objectives (process improvement goals and produce metrics).

Do what you say (compliance): follow all procedures and instructions that affect your work. You must say what you do (documentation): use current plans, procedures, and work instructions.

Prove it (records): demonstrate your work in accordance with sound processes/procedures and provide objective evidence.

Improve it: (process management/continual improvement): implement change based on information/metrics.

4.1.1 Organization-Level Policy

I feel that process models and quality management provide the necessary means for software organizations to establish effective software engineering review and audit processes to be implemented to support software development, modification, and software procurement programs. The number of processes and the extent of quality software engineering review and audit process implementation is based on the software program or project activities being performed each day. Software Quality Assurance (SQA) is an organization that will always ask audit questions concerning software development in order to support software engineering reviews and audit processes, because that is what software quality is allowed to do. At times this organization can be an annoyance to system, software, and test teams because of the many questions asked, including to subcontractors or supplier management.

The Software Configuration Management (SCM) organization and software quality team work well together in software activities such as audits, tracking of changes, software builds, loading of software in labs/sites, software supplier audits, verification/validation of processes for compliance, and software product development. I have been on both sides, and these organizations together make a very good team to ensure that software organizations or programs are compliant with required software standards.

Software development activities are performed in accordance with defined, repeatable, managed, and optimized policies. The software quality disciplines and software configuration principals ensure that software development processes are using a company or program standard, which shows that cost parameters are established, documented, and maintained. The peer review methods utilize that major software defects are addressed and will prevent future occurrences. In many software industries and military and aerospace programs, software cost estimates are used for standard estimation tools based on historical data and expert estimation techniques.

4.2 PLANNING PROCESS

To ensure that software plans will succeed, you must plan for a software engineering review and audit implementation. The software industries and military and aerospace programs require that reviews and audits are scheduled in the planning activity. To keep on course, allow members with quality experience and the knowledge of configuration management to identify milestones inside the planning process. I have witnessed software plans completed, and when it comes time to deliver to the customer, there were no specific or identified software engineering reviews and audits. What? No software engineering reviews and audits performed or scheduled in the software plan? How can the quality of deliverable software items under contract meet requirements when no verification, validation, reviews, or audits have been performed?

Software organizations or programs not included in the process of effective software planning fail. I have seen this happen before in many years of working in a software environment. Software industries and military and aerospace programs need to ensure that effective software planning processes are discussed prior to presentation for acceptance and implementation.

4.3 PLANNING PROCEDURES

Software planning procedures are prepared early in software development cycles to consider important elements of software design, testing, and deliveries to the customer. A software plan is critical to define the approach required to verify that all software items fulfill the correct requirements and needs. How can you be successful if you do not have an effective and sound software plan in place? When it comes time to evaluate a software plan for starting the activity inside a program, the first question is, do you have a software plan prepared and accepted by senior management along with other employees or team members?

I have been in planning sessions and discussions when an employee or team member presents a software plan to senior management and other team members for approval. It is interesting when senior management and the team members have not been made aware or seen the software plan being presented. That is not, and I repeat not, the way to kickoff a software plan. You must have software planning procedures created, developed, and approved by senior management before preparing and presenting the software plan. Taking responsibility for the implementation and execution of the required planning procedures will ensure that the software plans are successful.

4.4 CONTROL PROCEDURES

To establish contract requirements and assign responsibilities to the software teams for implementation, software control procedures are developed to implement policies and meet legal, regulatory, contractual, and operational requirements. Product specifications, assembly instructions, product standards, product definition documentation, or other technical process instructions that are governed by separate and independent infrastructures are not included in software control procedures. I will document the technical software disciplines required in this book, but let's talk first about the importance of having software control procedures in place. Without control procedures for software, how does a software organization meet the identified contractual requirements?

If a software company or military and aerospace program cannot meet contractual requirements or assign responsibilities to ensure compliance to software policies, you are in major, and I repeat major, trouble. Ensure that software control procedures are in place for the development of effective software plans, sound software processes, and implementation of critical contractual requirements.

FURTHER READING

Cassidy, Anita. *Practical Guide to Information Systems Strategic Planning*. Boca Raton, FL: CRC Press, 1998.
McConnell, S., *Software Estimation, Demystifying the Black Art*, Redmond, WA: Microsoft Press, 2006.

5

Systems Engineering

In context, systems engineering is associated with analysis, requirements understanding, and the importance of software design capabilities. Interfaces are defined externally and internally to ensure that hardware and software are compatible in supporting team activities. The systems engineering management plan provides direction for planning major technical reviews and includes configuration management for planning and preparation for formal audits.

5.1 SYSTEMS ENGINEERING METHOD

The systems engineering methods are included in tasks or assignments to integrate all system-related disciplines to meet system requirements for software industries and military and aerospace program technical needs. Methods used for systems engineering are also applications that set the ladder for rigorous software engineering techniques to solve complex problems, both technical and functional.

INCOSE defines systems engineering as "an interdisciplinary approach and means to enable the realization of successful systems."

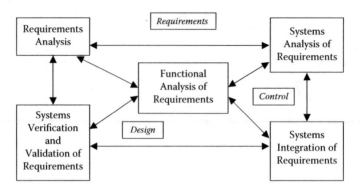

FIGURE 5.1
Systems engineering flow of requirements.

5.2 SYSTEMS ANALYSIS OF REQUIREMENTS

Effective systems analysis of requirements with technical planning and use of software design reviews will complement trade studies, planning, and disciplined implementation of requirements. The accurate and effective use of technical performance measurements provides a disciplined approach of software configuration management principles. Figure 5.1 provides the systems engineering flow of requirements.

5.2.1 Requirements Analysis

Requirements analysis is used to define software industries and military and aerospace needs, their mission, and the environment of critical operations. The development of systems engineering scenarios addresses the usage of the software development lifecycle. The systems engineering requirement analysis and software design constraints identify functional requirements for software development to maintain traceability to requirements.

5.2.2 Functional Analysis of Requirements

The functional analysis of requirements is the understanding of functions from the top level to the lower level derived by missing requirements. The functional software development lifecycle states and modes are established per system requirements. The timing, sequence, conditions, and probability of executing to define and redefine functional interface requirements apply to the system architecture.

5.2.3 Systems Integration of Requirements

The systems integration of requirements is the transformation of the functional architecture into optimal design solutions. The implementation of disciplined interface management principles, planning, resources of logical systems build-up, and integration activities is a well-defined delegation and execution of systems engineering requirements.

5.2.4 Systems Verification and Validation of Requirements

The robust approach for systems verification and validation of requirements is to plan, evaluate, and record software product compliance with designated requirements. Incremental risk reduction will assess and ensure that software design activities satisfy user needs and provide efficient and cost-effective integration of validation and verification of requirements.

5.3 SYSTEMS ENGINEERING MANAGEMENT PLAN

The Systems Engineering Management Plan (SEMP) establishes system-level technical software reviews conducted for military and aerospace programs. The major technical reviews affecting software are the system Initial Requirements Review (IRR), Interim Design Review (IDR), Final Design Review (FDR), Test Readiness Review (TRR), Functional Configuration Audits (FCA), and the Physical Configuration Audit (PCA).

The purpose of this plan is to address upgraded processes from a systems engineering point of view. The majority of the plan addresses and describes the software change management and authorization processes.

The plan for systems engineering is organized into three main sections: systems engineering, technical program processes, and engineering integration. The systems engineering section describes an orderly and structured approach to the overall system design, software development, and required formal reviews. A quality systems engineering management plan is a document that will provide software industries and military and aerospace programs with technical expertise to execute activities throughout the software lifecycle. Using the plan will enable performance to be more effective and productive, and technical planners will spend more time planning to ensure that customers have greater assurance in addressing technical challenges.

FURTHER READING

AS9100 Aero Space (AS) Standard Quality Management System Requirements—Guidelines for the Application, Part 2, 1997.

Department of the Air Force STSC, MIL-STD-499B System Approach for Systems Engineering of Defense Systems, Volume 1, February 1995.

SECAM: Systems Engineering Capability Assessment Model (Version 1.50A),

Seattle, WA: International Council on Systems Engineering , June 1996.

Wigers, K. E., *Software Requirements*, Second Edition, Redmond, WA: Microsoft Press, 2003.

6

Software Design

To improve your chances for a successful software design method, it is important to have a well-defined software process. You should define and document plans to support the software design methods. Software design methods and a documented plan for software development will help software developers. These methods provide the necessary support to organizations for them to know what is expected as far as software design/development activities. Appendix A provides a draft or outline that can be used to define and document a Software Development Plan (SDP). The critical items pertaining to the development of this plan consist of planning and providing engineering information and direction for the production of software. It is important to know that this planning process must be consistent with system-level planning.

6.1 SOFTWARE DESIGN METHOD

The purpose of the Software Design Method is to provide a systematic approach to the transformation of software requirements into design definitions for a software end item. The software architecture definition provides a framework for the creation of the software end item design and could provide constraints on the design. The resulting software design definition provides details about the software architecture, behavior, components, and interfaces. Established traceability between elements of the software design methods and the software requirements is critical to the software development process. The traceability data and software design definitions are documented according to software development plans and required processes. The required actions supporting the software design method are documented in the SDP.

This method is used for the creation of the software design and submittal of the software products for review and control. This method also applies to improving and providing corrections to the software design definition. A software design definition is created for each software end item. The method to the design of a software end item and lower-level components is established.

6.2 SOFTWARE DEVELOPMENT PROCESS

There are numerous definitions and questions in software design. What is the software development process, and how should this process be imposed? A software development process is a framework that will enable a software business or military and aerospace program to develop and deliver quality software. The common traits pertaining to the software development process are as follows:

- Whatever process is implemented, make sure it involves developing in iterations.
- Be sure to incorporate a way to evaluate how well the process is working.
- Incorporate approaches that will solve software issues and problems.

A good software development process will always deliver good software. A great software development process will allow your development team be successful. Many software teams love having a process in place. Software teams do change things, but they consider the following before that change occurs:

- Minimizing disruptive changes in the process
- Developing metrics to determine if a change is helping
- Valuing and respecting their fellow team members

It is important to focus on individual achievement, but it isn't easy to switch gears and emphasize team performance. But switch and emphasize teamwork while allowing for individual differences, and the whole team will win.

6.3 SOFTWARE DESIGN DECISIONS

The software architecture definition provides design concepts and constraints for the software end item design. The software requirements definition and the software operational concepts identify the capabilities and characteristics required of the software end item. These inputs are analyzed and integrated to make key design decisions for the software end item.

6.3.1 Evaluate Requirements and Architecture

Review the software requirements definition and software operational scenarios to ensure that issues affecting the software design are identified and resolved. Evaluate the feasibility and effectiveness of the software architecture definition as applied to the software end item. Review project resources and plans, and identify any change in scope that might impact the program and project. Deficiencies and conflicts in requirements, architecture, or plans will be reported to the stakeholders for resolution.

If reused or purchased software components have been identified by the software architecture definition, decide how these components will be incorporated into the software end item design. Identify additional opportunities for software reuse such as common underlying utilities and services. Use established reuse criteria as identified by software development plans to determine if the organization's reuse library and/or existing software work products can be used for potential software design reuse. Use trade studies and prototypes as needed to support make, buy, and reuse decisions.

6.3.2 Analyze Requirements and Architecture

Decide how the software functional and interface requirements will be transformed into a physical representation in the application domain. Decide how to apply the design intent and concepts provided by the software architecture definition to the end item architecture and design. Identify and understand how to apply design policies that guide the design such as abstraction, encapsulation, and information hiding. Develop additional design concepts that are specific to the architectural and detailed design of the software end item. Develop strategies and approaches as needed for the following:

- Software architecture frameworks and design patterns
- Operational concepts, states, and modes
- User interface concepts
- Communications concepts, interface standards, and protocols
- Data storage, access, integrity, and security
- Error handling and fault management
- Strategy for meeting critical requirements such as security and safety
- Strategy for meeting service and performance requirements

Evaluate alternative design solutions for major design decisions using trade studies, prototypes, models, or formal decision analysis methods, as appropriate. Provide motivations for major design decisions. Supplying the rationale for design decisions will preserve the integrity of the design through change. The design decisions and the rationales will be recorded according to the software development plans and defined processes.

The current software programs, software developers, or designers use a systematic and documented method for software development. These methods are described using a particular system architecture or software requirements, such as required subsystems and use of military standards (i.e., MIL-STD 480, MIL-STD 2167A, etc.) or the existing use of government/acquirer-furnished property (equipment, information, or software) as a requirement.

Military and aerospace programs use particular software design requirements or construction software standards, such as data standards, programming language, workmanship requirements, and production techniques. It is important to define and write a plan to support the software design methods. Successful software designers document a plan for software development, which will not only help the software developers or designers, but also will support systems engineering, test organizations, configuration management, and quality. It is important to know what is expected during software development activities. Software companies or industries should take a good look at utilizing military standards to help set up system architecture approaches and for implementing software requirements.

Some MIL-STDs are not available for use due to restricted use, but they could be reviewed and used as a guide for your software implementation.

6.4 COMPUTER APPLICATION METHODS

The computer application methods will identify and describe top-level software components, relationships, and behaviors. These sub-steps are iterated until all the software requirements and application methods have been addressed. In the business world, computer application methods are used in software design and development. I will address and provide a brief explanation of three (i.e., Waterfall, Spiral, and Agile) computer application methods.

6.4.1 Waterfall Software Method

The Waterfall Software Method is a single-step model. It applies to Department of Defense (DOD) Military and Aerospace Systems Development and Demonstration programs that are a single step to full capability, and can also apply to National Aeronautics and Space Administration (NASA) programs in software design phases. The software builds are incorporated to reduce risk, allow early user feedback, and support integration tests. But the general nature of the water software method requires that mature technology is used and that the performance requirements can be met with a "single step" to operational capability. Figure 6.1 shows a general Waterfall lifecycle model.

The Waterfall approach is where all software requirements are defined up front, then followed by design, development, and test to stabilize the iterative approach. Software development starts with the requirements of the desired outputs, such as software requirements or use cases for certain functions or features supporting use cases; design; and then software development. Notice in Figure 6.1 in the flow diagram box that *Determine Architectural Approach* is highlighted.

6.4.2 Spiral Software Method

The Spiral Software Method for a software development lifecycle will generally have a predetermined number of spirals specified or proposed within

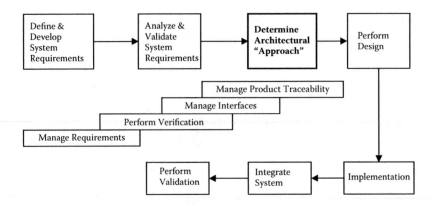

FIGURE 6.1
General Waterfall lifecycle model.

the increment under contract. At the beginning of each Spiral, the system performance requirements are refined and risk mitigation activities are planned. An example of a general Spiral lifecycle is shown in Figure 6.2. The overlap between spirals is notional and will depend upon the specific program plan. It is possible to have multiple software prototypes or software builds within each Spiral, although the intermediate software builds will not be formally delivered. They may be used to inquire user feedback and for risk reduction.

Within the software discipline, there are many elaborations and methodologies associated with the general Spiral development approach. Spiral Lifecycles are focused on project lifecycles, not specific software development lifecycles. Notice in Figure 6.2 in the flow diagram box that *Determine Architectural Solution* is highlighted. Having a solution versus an approach is basically the difference between Waterfall and Spiral methods.

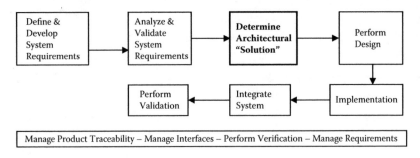

FIGURE 6.2
General Spiral lifecycle model.

6.5 DEFINITION OF AGILE

My definition of Agile is providing customer satisfaction through the rapid and continuous delivery of useful software. The working software is often delivered in weeks rather than months. It is important to note the working software measurements determine progress. Sometimes late changes in software requirements are okay, but have closure daily with cooperation between software designer/developers supporting the software industries and military and aerospace programs or projects. Face-to-face communication is a good form of conversation. Software projects are built around motivated software personnel who can be trusted. Continuous attention to technical excellence and achieving good software design is through

- Simplicity
- Self-organizing software teams
- Adapting to software changes

6.5.1 Agile Software Development Method

Software companies and military and aerospace programs that are lean and agile have a competitive advantage. By implementing Agile principles and practices, software development deliveries of products to customers will have fewer defects. Applying Agile methodologies to software development supports many initiatives and provides a project management approach that emphasizes short-term planning, risk mitigation, and adaptability to changing requirements, as well as close collaboration with software customers and the Integrated Product Team (IPT) to show accountability. Figure 6.3, showing an Agile management model, depicts the process.

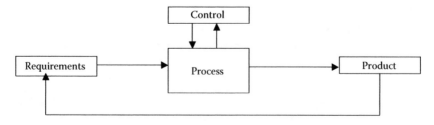

FIGURE 6.3
Agile management model.

6.6 DEFINITION OF LEAN

Lean is a new concept in the software world. Lean principles establish clear priorities by getting rid of bad multitasking, lack of focus, and not finishing the task assigned to an individual. Lean principles will eliminate the release of software late and require an early delivery. One must prepare, start, finish, and use checklists to prevent software defects and risk. Notice that I stated you must use checklists to support software engineering reviews and audits. Teams will face issues and resolve them in timely manner and drive daily software execution.

6.7 DEFINITION OF FIRMWARE

The definition of firmware has always been debated. Whenever I bring up the term "firmware" I get a confused look from software and hardware engineers. Software engineers say firmware is hardware, while hardware engineers say firmware is software. To me, the definition of firmware is an assembly composed of a hardware unit and a computer software program integrated to form a functional entity whose configuration cannot be altered during program execution. A computer program is stored in a hardware unit as an integrated circuit with a fixed configuration that will satisfy a specific software application or operational requirement. Does that make sense?

Another definition of firmware is as an embedded software program that facilitates the associated hardware in performing its required hardware functions. The relationship between the hardware and the firmware is considered to be interdependent, and the qualification testing of the hardware requires that the firmware be resident on the hardware.

As a quality assurance engineer, I support software loads, software upgrades, firmware (hardware component) upgrades, and hardware component replacements. I review and approve Software Load Requests (SLRs) to witness/verify software loads, software upgrades, firmware (hardware component) upgrades, or hardware component replacements (swaps) in accordance with installation procedures. The quality engineering function will perform the following:

- Participate in the appropriate software or functional testing to verify the success of implementation in accordance with procedures.
- Provide quality stamps on the software identification labels on the media.

Yes! We finally have a definition of "firmware." But the definition is still up for debate between software and hardware engineers.

Suppliers and subcontractors are required to be involved in the understanding of firmware. The software and hardware processes are documented to include standard engineering practices such as requirements development, verification and validation, peer reviews, and configuration management. The development of the software components of firmware follows the software-defined processes specified in a software plan. The development of the hardware components of firmware follows the hardware-defined process in the hardware plan. The next two paragraphs identify the software and hardware components of firmware. I hope this will put an end to this debate.

6.7.1 Software Components of Firmware

If executable code is firmware, then the firmware component is identified as software. The location where the executable code resides is identified as software. The requirements allocated to firmware are used to identify the software components of the firmware.

6.7.2 Hardware Components of Firmware

When firmware is a hardware configuration, the firmware is identified as hardware. The methods used to develop a hardware design include high-speed integration circuits. The hardware components of firmware will be identified, and system requirements allocated to firmware are used to identify the hardware components of the firmware.

6.8 IDENTIFY SOFTWARE COMPONENTS

The top-level software components of the software end item will be identified and described. Decompose the architecture into top-level design classes, components, functions, or objects, as specified by the design methodology, which are identified in the software development plans. Evaluate, select, and reuse available top-level software components, if applicable.

6.8.1 Define Relationships

The static relationships between top-level software components will be defined, such as inheritance and composition. In parallel, define other substeps, and in an iterative nature, evolve models or other design representations of the software components. Package diagrams, composite structure diagrams, component diagrams, class diagrams, or structure charts may represent static relationships. The architectural design representations may be preserved throughout the life of the software end item to aid communication with relevant stakeholders and to maintain the integrity of the design abstractions over time.

The dynamic relationships between top-level software components will be defined, such as timing, sequencing, synchronization, dependencies, and priorities. Evolve design representations to include behavioral relationships. Use case diagrams, sequence diagrams, interaction overview diagrams, communication diagrams or data/control flow diagrams to represent the behavioral relationships.

The concept of execution of the software end item and its components will be defined. Use task analyses, timing diagrams, deployment diagrams, state machine diagrams or other behavioral models to evolve the concept of execution. Describe the flow of execution control, handling of asynchronous and synchronous events, concurrent execution, and other applicable aspects of software end item execution.

6.8.2 Interfaces and Data

The external interfaces of the software end item and its components will be identified and described. Define external interfaces with hardware, users, off-the-shelf software, reused software, and other software end items. Class diagrams, structure diagrams, and block diagrams or data dictionaries may describe interfaces. Define internal interfaces between top-level software components at a level appropriate for the phase of the project. Identify interface standards and protocols. Define interface data types and formats at a level appropriate for the phase of the project. Develop data models, database designs, and data schema as needed.

6.9 DETAILED DESIGN

The detailed design is refined to the lowest level necessary for implementation. These substeps are iterated as needed to decompose the design from the level of software components to the level of software units in preparation for coding.

6.9.1 Components into Software Units

The top-level software components will be decomposed into lower-level software units. Use the concepts and constraints established by the architecture as guidelines to refine and expand the design to more detailed levels. Graphical representations may include class diagrams and structure charts. Evaluate, select, and reuse available software units for each software component.

Create detailed design representations using the methodologies and tools specified in the software development plans and defined processes. Define the design detail to the lowest level necessary to support implementation. Refine the detailed design until

- The design is specific enough to be implemented.
- The size of each software unit is maintainable.
- The number of interfaces is minimized.

6.9.2 Configuration Item Details

A Configuration Item (CI) is an identified aggregation of hardware, or any of its discrete portions, that satisfies an end-user function and is designated for configuration management. The computer software is uniquely identified as the aggregation of software, or any of its discrete portions, that satisfies an end-user function and is designated for software configuration management.

6.9.3 Interface and Data Design

The internal interfaces between software units will be identified and described according to the standards identified by the program or project. Refine definitions of external interfaces. Identify or design the mechanization of each interface. Define the protocol, communication method, data types, and data formats for internal and external interfaces. Graphical representations may include class diagrams and object diagrams. Interfaces may also be described in an Interface Definition Language (IDL) or Extensible Markup Language (EML). Interface details that are defined by existing standards or specifications will be referenced and not repeated.

Define the structure for databases, data stores, or data objects as needed. Graphical representations may include class diagrams, object diagrams, or entity-relationship diagrams. Define the logic for database operations, data manipulation, data collection, and data distribution. Evaluate, select, and reuse existing interfaces, data designs, and documentation, if applicable.

6.10 EVALUATE THE SOFTWARE DESIGN

The software design is continually evaluated as it is refined to confirm that it remains consistent with the intentions and constraints established by the software architecture definition. The software design is evaluated to ensure that it provides a complete and usable plan for implementation of the software requirements assigned to the software end item. The software design is evaluated against assigned quality attributes such as performance goals, reliability, extensibility, etc.

The software design is also evaluated against the reuse implementation plan(s) and the risk mitigation strategy. Prototyping, modeling, or simulations will validate the high-risk parts of the design, including reuse. Unresolved risks that may impact the program, including reuse risks, are forwarded to the program or project risk management system. The software design is evaluated against design quality criteria, including the following:

- Completeness at each level of abstraction
- Consistency from one level of abstraction to the next
- Minimized interdependencies between design elements (low coupling)
- Maximized logical consistency within a design element (high cohesion)
- Minimized complexity to what is essential to adequately solve the problem

6.11 SOFTWARE CODING AND UNIT TEST

The Software Coding and Unit Test tasks provide a systematic approach to the implementation of a configuration-managed software design to produce source code. Software coding also applies to the implementation of changes and corrections to the software product. The software designer/developer will ensure the appropriate level of unit testing to be performed and verify that the source code is ready for integration. Source code is tested against the design. For each unit tested, test procedures are developed and tests are run.

6.12 PEER REVIEW METHOD

Peer reviews are technical evaluations of work products (such as plans, specifications, trade studies, test procedures, and software code) for consistency and compliance with program or project standards and contractual requirements with the intent of identifying, preventing, and eliminating defects in work products early and efficiently. The evaluations, which are performed by peers with the knowledge and expertise to evaluate the work products, provide a gateway to the next development phase. When this process asset is selected for use on programs, the required actions are contained in the checklist. The peer review describes the planning for peer reviews, preparing and distributing the review package, conducting the review, and closing out the review. Table 6.1 summarizes the Peer Review Method and its interface requirements.

6.13 PEER REVIEW PARTICIPANTS

Peer review participants include knowledgeable representatives from all parties affected by the work product. This includes representation for the function that is producing the work product and for other functions that are technically involved in the work product. This representation may include subcontractors, teammates, partners, or other subject matter experts, including experts who are external to the program. Participants

TABLE 6.1

Peer Review Method

Input	Input to (Function)	Process Step	Output Required	Output Due to (Function)
List of products being reviewed	Program management and technical staff	Plan for peer review	Peer review plan	Review participants
Peer review plan, checklists, and guidelines; selected review forum; work product under review	Review participants	Prepare the work product for the review	Review package	Review participants
Review package	Quality, supplier management, suppliers, as applicable	Follow internal processes	Reviewers' comments	Review participants
Review package	Review participants	Conduct the review	Review package with identified errors, defects, action items	Review participants
Review package with identified errors, defects, action items	Review participants	Complete the review	Completed review package, accepted work product	Program/ project

typically do not include management or customers, unless they are technically involved with the work product under review.

Specific responsibilities are typically defined for review participants, including reviewer, presenter, and coordinator. Review participants fulfill at least one responsibility and may fulfill multiple responsibilities. Creators of programs or projects are encouraged to combine responsibilities to best fit their team member skills and their selected peer review forums, and to minimize unnecessary overhead. Typically,

- The presenter is the author of the work product.

- Three to eight participants have the reviewer responsibility.
- One coordinator is identified for all peer reviews or for peer reviews of one work product type.
- The presenter and coordinator responsibilities should not be combined because this represents a potential conflict of interest.
- Responsibilities may be further divided across participants (such as dividing the coordinator responsibilities for facilitating and recording).
- Review forums include, but are not limited to, face-to-face participant meetings, virtual meetings, or other electronic collaboration such as a peer review tracking system.

6.13.1 Peer Review Steps

Software peer reviews are held for work products and work product sections during development and revision. The precondition that triggers this process is the completion of a work product or work product section on which a program requires the performance of a peer review. In addition, the following preconditions will be defined as part of the Peer Review Plan: a schedule that includes review plans for new and revised work product sections and for a final, overall review at the completion of the entire work product as needed. The scheduled work product sections should be logically portioned for peer review to ensure manageable review packages. This schedule should show incremental peer reviews immediately after completion of each section and should not be delayed until the end of the phase when there is limited time for rework. Peer review guidelines include the following:

- Requests for appropriate functional participation for the various types of review work products.
- Participant responsibility assignments.
- Checklists for the work products of that phase to ensure complete and consistent review coverage for compliance with applicable standards and requirements. Checklists should be organized around defect categories and typically include checks for correctness, applicability, consistency, understandability, and completeness, and they address specific program priorities (such as speed, size, and maintainability). These checklists typically are not static, and recommendations for checklist updates may be made while executing the peer review process.

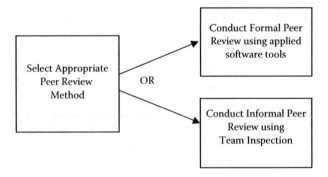

FIGURE 6.4
Peer review process steps.

The postconditions required for completion are that the review results are documented and distributed, action items have been tracked to closure, and any defects found have been documented in accordance with program or project plans.

A type of software review called a peer review could be considered a desk check, where another software designer will review the software work product. A sample of a formal and informal peer review process is shown in Figure 6.4. Comments from other software designers or developers during peer reviews are beneficial. The peer review responsibilities come from other software designers, testers, systems engineering, software configuration management, and software quality.

The peer review process is enhanced by adherence to uniform coding standards. The code is more accessible to potential reviewers, and less time is wasted adapting to different approaches. The review can focus on actual and potential defects and their causes. In addition to checking for adherence to standards, peer review leads will share ideas and improve coding techniques. Inspection by a software designer/developer prior to the peer review will contribute to defect prevention. Coding standards enable government audits of the peer review process. The review of software will be randomly reviewed on behalf of the customer in order to ensure that a uniform approach is being followed.

6.13.2 Establish Peer Review Plan

Establish plans, guidelines, and checklists to be used when conducting peer reviews such as when face-to-face versus electronic peer reviews will be conducted, what functional participation would be required, and participant responsibility assignments. Define the data to be collected for each peer review and establish the mechanism to record and store this data.

The peer review checklist for the work products will ensure complete and consistent review coverage for compliance with applicable standards and requirements. The checklist should be organized around defect categories and typically includes checks for correctness, applicability, consistency, understandability, and completeness, and it addresses specific program priorities (such as speed, size, and maintainability). These checklists typically are not static, and recommendations for checklist updates may be made while executing this peer review process.

Define the guidelines as needed to clarify any program- or project-specific entry, exit, and disposition criteria. Develop and maintain a schedule that includes review plans for new and revised work product sections and for a final, overall review upon the completion of the entire work product. The scheduled work product sections should be logically portioned for the peer review to ensure manageable review packages. The schedule should show incremental peer reviews immediately after completion and should not be delayed until the end of the phase when there is limited time for rework. The schedule should also include the dates for peer review training and for when materials for peer reviews will be available.

6.14 PREPARE THE PRODUCTS FOR THE PEER REVIEW

The following paragraphs define the peer review preparation activities that occur when a product or work product is to be reviewed and has been completed. These activities should be completed no less than two days prior to the scheduled review to allow for sufficient review time.

6.14.1 Prepare the Peer Review Package

Prior to preparing the peer review package, ensure that the work product is ready for review. Reschedule the review if the review package is not ready and update the peer review schedule. When the work package is ready, clearly identify the section of the work product for review and any appropriate supporting information. If this is a revised work product, ensure that changes are clearly marked.

Consider the size and complexity of the work product to be reviewed. If needed, partition the work product across multiple reviews. The ideal review work product size is about 20 pages and does not require more than one to two hours of review time. Using the work product checklist, review the work product and the supporting information to ensure that the information is sufficient and reasonable for the reviewer's needs. If needed, update the work product or provide additional explanations or instructions for review purposes.

6.14.2 Distribute the Peer Review Package

The peer review package will be prepared and distributed per the program or peer review guidelines. Ensure that the work products scheduled for review are being completed and submitted for review as scheduled. If needed, elevate any schedule issues to the team lead for resolution. Ensure that the work product, any supporting information, and the work product checklists are ready and accessible to the reviewers.

Select reviewers to provide coverage per the program's or project's participation guidelines and assign specific responsibilities. Consider any specific reviewer expertise required for this work product. If appropriate for the work product, include directed focus instructions for individual reviewers. Examples of focus instructions include specific areas of impact or interface (such as requirements, design, test, and operation/use), specific work product sections, or specific attributes (i.e., testability, usability, portability, etc.). Schedule the peer review forum. Consider other review and audit activities occurring on the program or project when scheduling to allow for adequate reviewer participation. Ensure the recording of preliminary information including review number, review date, time, place, invited reviewers, work product type, and work product size. Distribute the review package (i.e., review notice, work product to be reviewed, checklist, and any other supporting material) to the reviewers.

6.14.3 Conduct the Peer Review

In order to conduct the peer review, the requests for participation are distributed. Inform the distributor of the peer review package or the team lead if there are any issues with the participation requests. The success of the scheduled peer review is dependent on each reviewer's ability to provide a thorough review. The review package will be reviewed using the checklist for the work product.

Record comments identifying potential errors and defects. Also record any editorial comments or any recommended checklist updates. Record the amount of time spent reviewing the work product.

6.14.4 Categorize the Comments and Disposition

At the time of the scheduled peer review, ensure proper representation and preparation by the reviewers. Provide clarifications on the work products. Present comments and listen to the comments of the other reviewers. Comments can be presented either by page or by reviewer. Keep the comment discussions short with a focus on detection, not correction. Editorial comments are provided separately and are not discussed at the scheduled review.

Participate in categorizing comments. The comments will be categorized and documented as errors, defects, and action items. Refer to the definitions for the categorization rules, which are summarized as follows:

- Errors (i.e., problems in the material currently under peer review).
  - Optionally, errors are subcategorized as major (affects functionality and/or performance) and minor (does not affect functionality and/or performance).
- Defects (i.e., problems in materials previously peer reviewed).
  - Optionally, defects are also subcategorized as major and minor.
  - Note: Defects will further be categorized as delivered or nondelivered in the program's change request system.
- Action items (i.e., unresolved comments requiring further investigation)
- A comment can remain categorized as a comment if the reviewers and presenters agree that there is no error, defect, or action item required.

Assign responsibility for resolving the errors, defects, and action items. Ensure that the responsible parties are aware of this assignment. Record any discussion details that could be useful for future defect or causal analysis activities. This will include actions to update the review checklist.

Disposition of the review as complete, cancelled, or additional review is required. Reviews should be cancelled and rescheduled if the appropriate reviewers are not prepared or represented. An additional review should be scheduled for work products with a large number of detected errors after the errors are corrected.

6.14.5 Complete the Peer Review

To complete the peer review you must identify errors, defects, and action items to be resolved and documented. If needed, follow the program's or project's defined decision-making processes to elevate and reconcile any issues encountered in resolving peer review errors, defects, or action items with appropriate stakeholders. To ensure completion, perform the following:

- Correct all errors and update the peer review information to indicate that the error is resolved.
- Submit change request paperwork for all defects. The status and tracking of the defect corrections are then handled through the change request system. The defects associated with the peer review should indicate this transfer and are categorized as resolved, allowing the peer review to be closed.
- Resolve and complete all action items. If any action items cannot be completed within the two-week period, these action items should be moved to the program- or project-level action item tracking system. The action items associated with the peer review should indicate this transfer and are categorized as resolved, allowing the peer review to be closed.

6.14.6 Complete the Peer Review Records

Verify that all errors, defects, and action items have been properly addressed. Ensure that another review is scheduled and completed if the disposition of the current review indicates that a follow-up review is needed. Ensure that all data has been recorded and that the completed review package is appropriately stored. Recorded data is used for metrics analysis and includes work product size, number of reviewers, reviewers' preparation times, number of errors found, rework times, and resolution of all errors, defects, and action items.

Sign and date the peer review package and inform all participants that the peer review for the specific work product is complete. Submit the reviewed work products for configuration control in accordance with the program or project plans.

6.15 DESIGN TRACEABILITY

The design traceability between elements of the software design and requirements is maintained throughout the design process. The design traceability data will be documented according to the plans, processes, and software standards. The design is assigned including quality requirements for the design elements and interfaces of the architecture. The architectural design can be revised until all requirements are coordinated with one or more design elements. The design traceability will show that completed software design definitions cover all software requirements and that any design item can be traced to its own requirements.

6.16 DESIGN DEFINITION

The design definition for software will be documented according to the Software Development Plan (SDP), defined processes, and applicable standards. The design models using program- or project-defined standards, tools, and methods will be documented. The software standards will generate deliverable products according to required plans, processes, and standards. Record the previous developments in software development team records according to the SDP.

6.17 MEASUREMENT DATA

The measurement data generated during software design or development will be collected. Metrics identify the required measurements and associated issues where the measurements are generated.

6.18 CONFIGURATION CONTROL

Applicable software products from the software design method will be submitted for control in accordance with the Software Configuration Management Plan (SCMP).

6.19 SOFTWARE INSPECTIONS

A review is performed in accordance with the effectiveness of software design, quality processes, and techniques to ensure that the items to be delivered meet the approved software inspections. The software inspections are conducted on the software products in accordance with applicable program or project requirements. Each program may define additional requirements for the conduct of software inspections within applicable SDPs. Additionally, a software process evaluation may be planned to ensure process compliance. A Software Design Checklist is provided in Table 6.2.

6.20 SOFTWARE INTEGRATION AND TESTING

The purpose of software integration and testing is to provide programs or projects with a consistent approach for integration testing. Software units, components, and subsystems are assembled in accordance with documented software integration procedures. Software integration testing ensures that the elements are assembled properly according to the SDP. The level of software integration will determine if the assembled elements will be subject to verification and validation activities. This process of integrating elements, ensuring the elements are properly assembled, and subjecting the end items to testing is repeated in an iterative fashion until the entire software system or subsystem described has been assembled and tested.

TABLE 6.2

Software Design Checklist

No.	Description	Y/N/NA	Notes/Comments
1	Deficiencies and conflicts in requirements, architecture, or program/project plans will be reported.		
2	Design decisions and the decision rationales will be recorded according to plans and defined processes.		
3	Top-level software components of the software end item will be identified and described.		
4	Static relationships between top-level software components will be defined.		
5	Dynamic relationships between top-level software components will be defined.		
6	The concepts of execution of the software end item and its components will be defined.		
7	External interfaces of the software end item and its components will be identified and described.		
8	Top-level software components will be decomposed into lower-level software units.		
9	Internal interfaces between software units will be identified and described according to the standards identified by the project.		
10	Design traceability data will be documented according to plans, processes, and product standards.		
11	Design definitions will be documented according to plans, defined processes, and standards.		
12	Measurement and estimated data will be collected.		
13	Applicable work products will be submitted for peer reviews in accordance with program or project plans.		
14	Applicable work products will be submitted for control in accordance with program or project plans.		

6.21 SOFTWARE FORMAL TESTING

Software formal testing provides a consistent approach to verify that software requirements have been implemented to required standards. The results of the test activities from test planning through test completion will include analysis and documentation. In this context, regression testing is just one aspect of software formal testing and should be accounted for in the test plans.

6.22 SOFTWARE USER DOCUMENTATION

Software user documentation will provide the approach for capturing instructions necessary to install, use, and maintain the software product in its target environment. It identifies the work necessary to develop user documentation associated with a specific version of a released software end item according to program or project plans, schedules, and requirements.

6.23 MAINTAIN SOFTWARE DEVELOPMENT

In order to maintain software development, identification of the work is required in order to update and maintain released software. Software maintenance planning, implementation, and testing begin when the list of Software Change Requests (SCR) for an update has been approved.

FURTHER READING

Arlow, J., *UML 2 and the Unified Process Practical Object-Oriented Analysis and Design.* Upper Saddle River, NJ: Addison-Wesley, Pearson Education , Rights and Contracts Department,

Arlow, J. and Neustadt, I., *UML 2 and the Unified Process, 2nd edition, Practical Object-Oriented Analysis and Design,* Upper Saddle River, NJ: Addison-Wesley, Pearson Education, Rights and Contracts Department, 2005.

Floras, W. A., and Carleton, A.D., *Measuring the Software Process, Statistical Process Control for Process Improvement,* Indianapolis, IN: Addison-Wesley, Pearson Education Corporate Sales Division, 1999.

Jameson, K., *Multi-Platform Code Management*. ISA Corp., Sebastopol, CA: O'Reilly & Associates, 1994.

Koch, G., *Oracle7: Tthe Complete Reference, Versions 6 & 7*, Berkeley, CA: Osborne McGraw-Hill, 1993.

Pilone, D., and Miles, R.. Head First Software Development. *Software Development*, Sebastopol, CA: O'Reilly Media, 2008.

Schwaber, K,, and Beedle, M.. *Agile Software Development with SCRUM*, 2002, Upper Saddle River, NJ: Prentice Hall, , 2001.

Putnam, L.H., and Myers, W., *IEEE Computer Society, Executive Briefing, Controlling Software Development*, Los Alamitos, CA: IEEE Computer Society Press, 1996.

Spiewak, R., MITRE Corp., and K. McRichie, Galorath Inc. MIL-STD 498, Software Development and Documentation. *CrossTalk, Journal of Defense Software Engineering*, Hill AFB, UT, 2008.

DID-SW-005: Data Item Description (SDP)

MIL-STD 2197A: Software Development and Documentation.

MIL-STD-480: Configuration Control: Engineering Changes, Deviations, and Waivers.

7

Software Quality

The purpose of the software quality is to provide a common operating framework in which best practices, process improvements, and cost avoidance activities can be shared and quality assurance responsibilities can be assigned. The expected results from converging shared quality best practices are improved process execution and reduction of operational costs. To be internally consistent, the definition Software Quality Assurance (SQA) will be used throughout this book. This definition does not in any way prevent the applicability of software programs known by other names, such as Software Quality Engineering (SQE) or Engineering Quality Assurance (EQA).

7.1 SQA AND SQE/EQA TERMINOLOGY

With the advent and incorporation of Capability Maturity Model Integration, or CMMI, the term "Process and Product Quality Assurance" (PPQA) is applied. Software Quality Assurance (SQA) which specifically addresses software engineering is the more generic term. Software Quality Engineering (SQE) and Engineering Quality Assurance (EQA) address both systems and software engineering.

TABLE 7.1

Software Quality Activities

Input (Product/Service):	Output (Product/Service):
Request for proposal contracts	Noncompliances
Supplier Data Requirements List (SDRL) items	Audit records
Software Program definition	Evaluation records
Work product under review	SQA engineering plans
Requirements definition	SQA activity records
Design definitions	
Verification of configuration records	
Verification plans	
Verification of procedures	
Verification results	
From (Supplier):	**To (Customer):**
Program management staff	Program management

7.2 SOFTWARE QUALITY METHODS

SQA personnel support Integrated Product Teams (IPTs) by encouraging a cooperative, proactive approach to software quality. SQA ensures process compliance with the Software Development Plan (SDP), and each product SDP, through evaluations and subcontract management participation. The results are reported to management.

For purposes of clarity and specific direction, the site instructions referenced herein shall be used to conduct SQA activities.

SQA, as part of the IPT, is responsible for ensuring compliance with the contract and adherence to software plans (e.g., SDP, SCM, SEMP, etc.). Each SQA engineer will be trained on the methods and tools needed to perform evaluations and audits. Compliance verification is performed using process evaluations, assessments, reviews, or appraisals. SQA witnesses/monitors test in accordance with the project-level SDP and are also active participants in project-level reviews and meetings.

In addition, the term Software Quality Assurance Plan (SQAP) will be used throughout this book, but does not prevent the use of other similar planning documents known by other names. This book defines the basic activities required for SQA support of software programs, as shown in Table 7.1.

FIGURE 7.1
Site-level program/project organization chart.

7.3 SOFTWARE QUALITY ORGANIZATION

A software quality organization is shown in Figure 7.1. The SQA Rep has a reporting channel to the software quality manager who then reports to senior management independently of program/project management and engineering organizations. This independence provides an important and needed strength to the SQA organization, allowing SQA to provide senior management with objective insight into the programs. SQA provides senior management the confidence that objective information on the processes and products are being evaluated and reported.

SQA has the resources, responsibility, authority, and organizational freedom to conduct and report results of objective evaluations/audits and, when necessary, initiate and verify corrective actions. The program or project SQA implements the quality plans at the project level in a manner that reflects site, program, and contract requirements. SQA verifies products and processes for compliance with criteria specified in approved program and project planning documentation.

7.4 FUNCTIONAL SOFTWARE QUALITY TASKS

The coordination and implementation of program or project plans will be performed with responsible program software quality engineers who will review and approve any tailoring or waivers of quality processes implemented by software programs or projects. Support is provided to the Process Action Team (PAT) initiatives and activities for common software

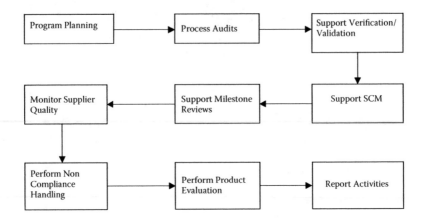

FIGURE 7.2
Major SQA process tasks.

process implementation, measurement, and improvement. Support of supplier activities for software is provided by the quality organization. Refer to Figure 7.2, major SQA process tasks.

7.5 SOFTWARE QUALITY PROGRAM PLANNING

Software quality program planning begins as early as possible in the proposal phase of the software development program or project. SQA participates in intergroup coordination between the proposal team, the software group, and the overall quality function for new and ongoing software programs. SQA planning and budgeting is an integral part of all software development programs from the onset. Late involvement of SQA in the software program should be avoided. Qualified SQA personnel or SQA expertise is used to plan and ensure adequate funding, and resources are set in place to properly sustain the SQA program plan. Proposals include first-time proposal programs and updates to the budget cycle of ongoing software programs. Figure 7.3 identifies the flow of the SQA program planning process.

7.6 PROCESS AUDIT

The SQA Process Audit Checklist in Table 7.2 is used to conduct process audits of software processes. SQA conducts these audits at planned

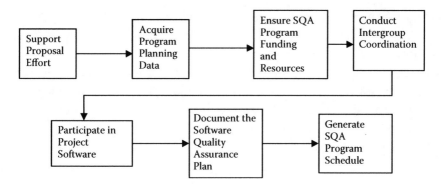

FIGURE 7.3
SQA program planning process flow.

TABLE 7.2

SQA Process Audit Checklist

Item	Description
1	Audit scope and criteria will be defined from previous audit history, organization procedures, project-level development plans, or other documented requirements.
2	Audits will be scheduled based on project schedules and activities.
3	Results will be recorded, including any noncompliances or observations.
4	An audit report will be generated that contains, at a minimum, area audited, scope and purpose of the audit, completed checklists, audit criteria, participants, results, noncompliances/observations, and any lessons learned for future improvement.
5	Audit reports will be approved by SQA management or its designee and distributed to relevant stakeholders.
6	Noncompliances will be addressed.
7	Actual measurement data generated during the conduct of this method will be collected.
8	Applicable work products from this procedure will be submitted for control in accordance with software development plans.

intervals and as needed during the software lifecycle to verify conformance to organization- and project-tailored procedures.

7.7 SUPPORT VERIFICATION AND VALIDATION

This paragraph describes SQA activities performed in support of software verification and validation. The purpose of software verification is to ensure that the software product meets the requirements, which is accomplished through various activities ranging from in-process Peer Reviews (PRs) and unit-level testing to software qualification activities, which include testing, inspection, analysis, and demonstration. The purpose of software validation is to ensure that the software products fulfill their intended use when placed in their intended environment. SQA will audit unit or other lower level test activities and perform software qualification activities per the process audit.

7.8 SUPPORT SOFTWARE CONFIGURATION MANAGEMENT

SQA activities are used to support Software Configuration Management (SCM) processes. These activities are performed in addition to the normal process and product audits and include participation in software review/change control boards as planned, software master verification, and configuration verification for delivery. To satisfy configuration verification for delivery, SQA verifies the completion of software configuration verification audits, or their equivalent, prior to the delivery of a contractual software end item. The audit provides evidence of as-designed/as-built compliance of the deliverable item.

7.9 SUPPORT MILESTONE REVIEWS

SQA verifies entry and exit criteria for milestone reviews based on contractual requirements and program standards. SQA participates in milestone reviews to the extent required, and the focus is on the SQA activities required to support milestone reviews.

7.10 MONITOR SUPPLIER QUALITY

The SQA activities are used to monitor software supplier quality management. This activity is applicable to suppliers that are delivering software that is developed or modified for software programs, whether delivered separately or embedded as part of a system (firmware or hardware). There has always been a debate or discussion about firmware. Is firmware software or hardware?

7.11 SOFTWARE NONCOMPLIANCE HANDLING

SQA initiates a closed-loop noncompliance handling system when SQA or other members of the program or project identify noncompliance issues. Immediate action addresses the removal of observed noncompliance, whereas long-term action addresses the root-cause activities to prevent recurrence. Preventive actions are performed to prevent a potential noncompliance issue from occurring. Noncompliance issues can be identified through various activities, including engineering peer reviews, testing, and SQA process audits or final product evaluations. As a result, noncompliance handling will be addressed through an engineering problem report or be documented in a corrective action system.

This paragraph addresses the process for SQA noncompliance handling. Implementation of this process will include product evaluations scheduled inside software organizations or programs. Waivers for noncompliance issues may be approved only through an established process, or by Senior Management or Software Quality Engineering Management, and must be documented.

7.12 PERFORM SOFTWARE PRODUCT EVALUATIONS

Software Product Evaluations are used to provide software programs or projects with the process steps necessary to conduct evaluations of software products. Software product evaluations ensure that the software product meets its specified requirements and are accomplished through various activities including in-process Peer Reviews. Peer Reviews may be performed and documented in accordance with the program's or project's Peer Review process. Required actions for Software Product Evaluations are contained in the following descriptions:

- Audit scope and criteria will be defined from previous audit history, organization procedures, project-level development plans, or other documented requirements.
- Audits will be scheduled based on project schedules and activities.
- Results will be recorded, including any noncompliances or observations.
- An audit report will be generated that contains, at a minimum, area audited, scope and purpose of the audit, completed checklists, audit criteria, participants, results, noncompliances/observations, and any lessons learned for future improvement.
- Audit reports will be approved by SQA management or a designee and distributed to relevant stakeholders.
- Noncompliances will be addressed.
- Actual measurement data generated during the conduct of this method will be collected and noncompliances will be addressed.

During the software development phase, contractual product evaluations performed on deliverable software products ensure compliance will be

- Based on contract requirements, business unit requirements, software development milestones, and project level plans, ensure that software products that require evaluations are identified.
- Software product evaluation criteria and plans are established based on contract requirements, business unit requirements, software development milestones, and project level plans.
- Software product evaluation records are documented and identify the product under review, evaluation criteria, participants, results, and findings. Identify any lessons learned and propose process improvements that will improve the software development and evaluation process.
- Resolve findings and elevate unresolved issues as needed. This will occur prior to customer delivery.

7.13 SQA ROLE IN AGILE SOFTWARE DEVELOPMENT

The SQA role in Agile software development is documented here. This paragraph provides detailed information for software industries and military and aerospace programs or projects. The Agile software development method is progressing and will replace current software development activities in the future.

SQA is "independent" and provides oversight, verification, witnesses, and support validations of the required software development. The actual software work is not developed by the quality team, but encourages the right balance between traditional quality roles and the Agile software team concepts. Early quality team involvement supports the identification of key Agile member involvement in the review of important program or project artifacts. The quality team encourages active communications between product owners and customers, but makes sure clearly documented and healthy Agile practices are institutionalized. Always watch for indicators so that software development is not reverting back to the Waterfall or Spiral software methods.

The Agile software team defines what it means to be "complete" with the implementation of software requirements. The quality team's role in software development provides independent reviews, evaluations, and interpretation of software quality metrics. It helps when the quality team can drive early measures of software quality. The Agile software team will identify corrective actions, process improvements, and report the results to the quality team. In order to support the preparation of defect prevention and process improvement plans, the Agile team with the quality team's support will

- Trace and verify requirements.
- Implement requirements to test cases.
- Review Version Description Document (VDD) baseline contents.
- Participate in and provide quality inputs in Agile review meetings.
- Review lessons learned and best practices from other Agile teams.
- Drive early measures of software quality assurance.

What is interesting is that quality is applied and consistent, per Figure 7.4, quality team's support to multiple Agile teams. Quality is embedded in the Agile software movement.

7.14 SOFTWARE QUALITY ASSURANCE PLAN

Documenting a Software Quality Assurance Plan (SQAP) will provide the necessary support and direction for how software organizations and

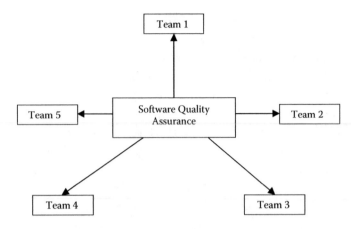

FIGURE 7.4
Quality support to multiple Agile teams.

military and aerospace programs should perform software quality assurance. In military and aerospace programs, I have been given the opportunity to define, document, and release this plan. Software programs should look at documenting an SQAP to define the quality of the software products released to customers.

7.15 SUMMARY

Software Engineering—Quality has a major task in Software Engineering Reviews and Audits. I would like to see quality teams step up, take charge, and be responsible in preparing and performing these reviews and audits. What it takes is being involved with software IPTs every day of the week or months. I know the quality team's role is to be independent, but get ingrained and you will see confidence and support from software IPTs. I attend every software meeting and verify software builds and software loading in Software Integration Labs and on all the aircraft and planes. The military and aerospace program management, systems engineering, software IPTs, testing, and software leads respect me due to my involvement. It is time for Software Engineering—Quality to make this change. Thank you for listening to me.

FURTHER READING

Kandt, R.K., . *Software Engineering Quality Practices*. Boca Raton, FL: Taylor & Francis, Auerbach Publications, 2006.

Russo, C. W. Russ, *ISO 9000 and Malcolm Baldridge in Training and Education, A Practical Application Guide*, Lawrence, KS: Charro Publishers,1995.

Zahran, Sami. *Software Process Improvement Practical Guidelines for Business Success*, Edinburgh Gate Harlow, UK: Addison-Wesley, Pearson Education, 2008.

Shoemaker, D., *A Comparison of the Software Assurance Common Body of Knowledge to Common Curricular Standard,*. Software Engineering Education and Training 20th Conference , Dublin, Ireland, 2007.

8

Software Configuration Management

Before I talk about Software Configuration Management (SCM), I want readers to know of my passion for being an SCM engineer. I was a software developer early in my career and enjoyed being a software engineer, but working in SCM gave me such a perspective of how important it is to have software configured, controlled, built, and then deliver the software to be tested and then provide quality software to the customer for implementation.

This chapter describes the organization of SCM and the practices applied consistently throughout the software lifecycle for Computer Software Configuration Items (CSCIs) that are developed and maintained by software programs. SCM focuses on identifying and managing changes and maintaining the visibility of the configuration of the software and its documentation. SCM is a cross-functional process applied over the lifecycle of a product and provides visibility and control over the product's functional and physical attributes. SCM processes are used during all phases of software development, from software program initiation to software product delivery, providing disciplines that identify the software products, establish and control software baselines, and document and track changes to the software baselines. These SCM processes control the storage, access, changes, archive, and release of the software products.

8.1 SOFTWARE CONFIGURATION MANAGEMENT METHODS

SCM methods are used during software development and maintenance of computer software. There are many SCM methods used by many software programs worldwide. The terminology, definitions, or terms used by SCM can be different, depending on the software program with which you work.

Examples include the following: an Engineering Review Board (ERB) is also called an Engineering Review Board (ERB), and a Software Change Request (SCR) can be used on other software programs while the term Software Change Request (SCR) is defined. Another definition is Version Description Document (VDD). Another definition is the Version Description Document (VDD). The (VDD) provides a detailed description of the software contained within an engineering environment, as well as software builds, test, versions, configurations, and file information. I have seen it throughout my 25 years working in military and aerospace programs. It comes down to this, that regardless of what is defined in the program or project, SCM will determine what definitions or terms referred to while working on a software program.

As a software designer for many years, I was comfortable to know that my software design activities were important and to release my code for SCM to configure and control, along with building my software code changes or updates to release for testing, validation, and delivery to my customers. There is more about SCM organizations that software designers do not know. Many software designers feel that SCM has no understanding of software engineering; I beg to differ—there are numerous software designers who want to advance to the next position available, and that is SCM, Test, or SQA.

The SCM activities that follow are necessary to do the following:

- Establish and maintain the (CSCI) identification process and control changes to identified software products and their related documentation.
- Record and report information needed to manage software products effectively, including the status of proposed changes and the implementation status of approved changes.
- Maintain auditable records of all applicable software products that will help verify conformance to specifications, interface control documents, contract requirements, and as-built software configurations.

8.2 ORGANIZATION

The SCM organization is represented within all the software IPTs independent of the software design representatives. The SCM organization is responsible for maintaining configuration control over the software developmental configurations and baselines, processing software changes,

Computer Program Library (CPL) operation, software construction, and management of the software configuration management tools.

The responsibility for SCM is assigned to the Software Manager along with the members of the SCM team. The disciplines are implemented throughout a program's software development lifecycle. Items for software are stored within the CPL while in the phases of software development, release to testing, and deliveries to the lab and customers. The SCM group has the primary responsibility for software configuration management. Personnel are full-time members for the duration of the software contract and participate in all software development, integration and testing, and software product release activities.

8.2.1 Software Build Engineer

The Software Build Engineer will perform the tasks related to software construction and configuration control, including the following: creation of a build folder to store the documentation of the build, source code changes, and records of computer program development. Software Build Requests (SBRs) are written and a build procedure checklist is provided to assemble, compile, and link source code; build archive copies; provide listings for use in software development, testing, and delivery support; and document the Version Description Document (VDD).

8.2.2 Software Change Request

The Software Change Request (SCR) coordinator will maintain the tracking system to control the software code and software documentation change status. The coordinator will administer the change database, coordinate, provide software inputs to program changes, and establish traceability to higher-level change authority impacting software. Another responsibility is to interface with Business Operations in tracking all software schedule commitments. Plans are reviewed to coordinate the ERB maintain of records including status of problem reports, changes, and document releases.

8.2.3 Software Release

The Software Release Coordinator will be responsible for the identification, release, and distribution of software and related documentation. The Software Release Coordinator's responsibilities include the following:

- Administration of the identification system for each software end item
- Coordination of software documentation releases through Data Management (DM)
- Providing documentation that reflects the released configuration of each CI/CSCI
- Issuing copies of unreleased software to other organizations
- Maintaining distribution records of drawings and documents

8.2.4 Engineering Review Board

The Engineering Review Board (ERB) is established for the software Integrated Product Team (IPT) to review and disposition changes that affect the controlled software and related documentation. All software changes will be documented, approved, and implemented under an SCR. ERB meetings will be scheduled and coordinated by the SCM IPT representative. The IPT leader or designated representative will serve as the ERB Chairman. ERB members will include, at a minimum, the following: the SCM representative, representatives of the affected software IPTs, the software testing representative, the system engineering representative, the quality representative, the security representative (if change impacts classified or trusted software), and the change sponsor. The major activities performed by the ERB are the evaluation and disposition of SCRs, assignment of priorities and action items, review of action items, change disposition at prior meetings, and the evaluation of deviations.

8.2.5 Configuration Management Plan

The Team CMP is described in this plan and includes the plan for software configuration management of software items developed during the program.

8.2.6 Software Development Plan

The SDP establishes the plan for the development of software during the design phase of the program. This plan establishes system-level engineering standards, practices, and guidelines for the development of CI and non-CI system software. The SDP coordinates with the CMP by describing the requirements to build in the configuration control of software as a function of the software development process, and by defining the interface between software developmental control and the formal configuration management control defined in the CMP.

The SDPs establish the software development plan for software developed by suppliers during the development phases. These plans tailor the SDP to reflect the engineering standards, practices, and guidelines utilized by suppliers for the development of deliverable and nondeliverable software items. SDPs will coordinate with the CMP by identifying the techniques to be applied in incorporating software configuration control in the software development cycle, in compliance with program configuration management control defined under the CMPs.

8.2.7 Operating Procedures

The SCM organization will develop and enforce SCM operating procedures, as identified under the description of the implementation of processes required to satisfy the requirements and the direction provided under the CMPs and SDPs identified above. Refer to Software Configuration Management Plan, Table 8.1.

8.2.8 Computer Program Library

The Computer Program Library (CPL) is established to preserve and control software media, documents, and associated management records. The CPL is a repository for all controlled software products and consists of both off-line storage and on-line storage on the Software Engineering Environment (SEE). The CPL will be maintained for the duration of the program. The CPL provides a controlled link between software development personnel and released software. Software developers are allowed access to data while ensuring that the software development configuration is maintained. This includes the ability to reconstruct or retrieve previous software versions. The Computer Program Librarian will maintain the controlled CPL. The librarian is responsible for the storage of source, object, and executable code and listings; maintaining records of computer program development status, including a Software Build Request (SBR) log, software media log, and CPL Inventory log; and maintaining copies of all released documentation, including release orders, vendor-furnished media, and Version Description Documents (VDDs).

8.2.9 Computer Program Library Contents

CPL data generated by SCM includes records, reports, and forms required to maintain control over the externally generated data submitted to the CPL. CPL data originated externally is primarily code, documentation, and supporting test data. All data types may be stored electronically or on hard copy.

8.2.10 Configuration Control

The CPL is controlled and maintained by SCM to provide for the integrity and reproducibility of software and associated documentation. Control over products residing in the CPL is maintained as follows:

- Physical access to the off-line CPL is limited to SCM personnel.
- After data or documentation is submitted to the CPL, changes cannot be made directly to the data.
- Changes are made to copies of the data and submitted with the appropriate authorizing change documentation to SCM for inclusion in the CPL.
- All changes to data contained in the CPL are authorized by an SCR.
- Records of software development status change tracking and release coordination are maintained in the CPL.
- CPL security is maintained per contractually imposed team security standards and the security manual.
- Historical records of all deviations and waivers applicable to software are maintained.

8.2.11 Developmental Configuration

The Developmental Configuration is an evolving CI-level configuration that is subject to increasing levels of configuration control as the software and associated technical documentation are developed. The Developmental Configuration is first established upon successful completion of Software Specification Review (SSR) reviews and consists of the SDP, CMP, and preliminary versions of the Software Requirements Specification (SRS), and Interface Requirements Specification (IRS) documentation. The Developmental Configuration evolves as the program progresses and eventually consists of the above-mentioned plans, updated

SRS and IRS, a Software Design Description (SDD), source code listings, and the changes that have accumulated since the last formal release of the CSCI. The data identifying changes to the CI is collected in Software Development Files (SDFs) until it can be formally released as an attachment to the software product specification.

The following design reviews and software configuration audits represent the major milestones under the Developmental Configuration:

- Software Specification Review
- Software Design Review
- Preliminary Design Review
- Critical Design Review
- Test Readiness Review
- Final Design Review
- First Article Inspection
- Functional Configuration Audit
- Physical Configuration Audit

8.3 CONFIGURATION IDENTIFICATION

The Configuration Identification is supplemented for software technical documentation and code through the establishment of the Developmental Configuration, defined under DOD-STD 2167A and tailored by the Software Development Plan (SDP). The Configuration Identification of software items and related documentation will include the identification of configuration baselines and configuration specifications, as well as assignment of CIs, Nondeliverable Items (NDIs), Computer Software Components (CSCs), Computer Software Units (CSUs), and Part Numbers.

A CI represents a group of CSCs defined for the customer for purposes of technical understanding, system capability definition, or functionality. Each CSCI is identified with a unique name, acronym, and identification number. CSCI numbers will be assigned. The files that make up each CI will be stored as a separate design part in the CPL online library. The SBR requesting the CI build will identify the CSCs and CSUs by filename, version, and baselines that comprise the CIs.

8.3.1 Computer Software Component Identification

The Computer Software Components (CSCs) are a logical or functional grouping of CSUs to which the SCM tools assign a unique name, which is supplied by the software designer. Specific CSC names and contents are defined in the Software Design Document (SDD). The SCM tool adds a version number to the end of a CSC name for modification accountability. The SCM tool will maintain information on the CSC modification history, authorizing change number, and user name.

8.3.2 Computer Software Unit Identification

Computer Software Unit (CSU) file names will consist of a three-character CSCI abbreviation, followed by a three-character CSC abbreviation and a descriptive CSU name (abbreviated to meet operating system and character limits) plus a version number. SCM tools maintain information on the CSU modification history, authorized change number, user name, version number, date, and time the file was modified.

8.3.3 Nondeliverable Software Identification

Nondeliverable Software utilized for development, maintenance, or operation support will be identified as a Nondeliverable Item (NDI). The files that make up NDI software will be stored in the SCM tools as separate design parts. The software is stored, assigned, and labeled on the media to reflect the software release and version.

8.3.4 Nondeliverable Software Control

Nondeliverable Software Control will be used to assist in the development of the deliverable CSCIs, but is not identified as a deliverable product. Control of the Developmental Configuration will be maintained using the lower- and middle-level controls of products depending on the maturity level and release status of the software involved.

8.3.5 Part Number Identification

Each configuration of a CSCI or NDI is assigned a unique part number. The part number will be assigned in compliance program or project standards.

The part-numbering scheme will only apply to software products at the CSCI/NDI level. Any unique identification at the CSC or CSU level will be accomplished with the use of mnemonics within the SCM product(s).

8.4 CHANGE CONTROL

Software products defined under a Developmental Configuration are placed under increasing levels of configuration control. These levels of change control determine the type and level of authorization required before changes may be implemented into the software product or associated documentation. Three levels of configuration control are applied to software products in the Developmental Configurations.

8.4.1 Lower-Level Changes

Software products defined under a developmental configuration are placed under increasing levels of lower-level configuration control. These levels of control determine the type and level of authorization required before changes may be implemented into the software product or associated documentation. Lower-level configuration control is applied to software products in the Developmental Configurations.

8.4.2 Middle-Level Changes

Middle-level change control is applied to products that have been released, but have not been approved. Changes to these products require team coordination for the evaluation of impacts to a released product. Changes to controlled products require an approved Middle Change, at a minimum. These changes will be documented via an SCR and coordinated within the team utilizing the change documentation (i.e., Change Request). This will require coordination among the SCM organization and the Team Change Management organizations for proper change-processing support.

8.4.3 High-Level Changes

High-level change control is applied to products that have approved change items coordinated with affected IPTs. Changes to products under high-level change controls are documented via SCR, processed, and authorized under high-level change control.

To supplement the change process, SCRs will be used to document and implement all changes to software products, allowing the SCM organization the ability to maintain the change status.

8.5 CONFIGURATION STATUS ACCOUNTING

Configuration Status Accounting contains records of detailed data that documents that the as-built software conforms to its technical description and its specified configuration. This data includes records of approved technical documentation for each CI/CSCI, implementation status of approved changes, software problems, associated status, and a log of SCR status. These SCM configuration status accounting records are maintained and used to supplement the configuration accounting records.

8.5.1 Software Records

The SCM organization maintains three categories of configuration records as electronic media with variable formats developed, as needed. These categories include software records, release and distribution records, and archive records.

Software records consist of the SCR reports, SBR reports, SDFs, code baselines, change closures, software build logs, and the CPL inventory log. The SCR reports provide a record of all SCRs and related information pertaining to any given CI or NDI. This information is maintained within the SCM databases and is generated as a report, as required. The SBR log provides a record of all software updates submitted by SBRs. The CSCI/NDI software build log provides a record of the events that took place during the SCM build, which is a supplement to the build information created by the SCM personnel performing the build. The CPL inventory log contains a record of each software product stored in the CPL, either electronically or in the physical CPL area.

8.5.2 Archive Software Records

A record of all archived material will be maintained by the SCM organization. Archived material includes obsolete material, data not required for current use, and off-site stored backup data for use in case of loss of on-line data.

8.5.3 Reports

The SCM organization is responsible for managing, compiling, maintaining, and publishing the software configuration accounting reports. These reports, produced on demand, provide evidence to management that all changes to the software are being accounted for in a one-to-one relationship. This evidence, together with the configuration status accounting reports maintained by the Configuration Management organization, are input to the customer for product acceptance.

8.5.4 Test Configuration

The management of software end items under testing will be controlled utilizing required software tools. SCM will establish and uniquely identify test configurations beginning with the first occurrence of CSC integration testing and continuing through Qualification Testing and system integration. SCM will perform the software builds and control the software under testing using SCM functions of problem reporting, baseline management, and reports on SCRs incorporated in the test configurations.

8.6 SOFTWARE CONFIGURATION AUDIT CHECKLIST

The Software Configuration Audit Checklist is provided in Table 8.1.

TABLE 8.1

Software Configuration Audit Checklist

No.	Description	Y/N/NA	Notes/Comments
1	SCM activities are defined in the SCMP.		
2	Software Products are identified.		
3	Configuration Identification is defined.		
4	Computer Program Library is defined.		
5	Computer Program Library provides storage, retrieval, configuration control, and disaster recovery.		
6	Configuration Control is defined.		
7	Software Products are controlled.		
8	Performing Software Builds is defined.		
9	ERB responsibility and authority are defined.		
10	ERB reviews the SCRs and approves, requires more analysis, holds, or rejects.		
11	ERB conduct will be recorded and maintained.		
12	Configuration Status Accounting is defined.		
13	Software Configuration Audits are defined.		
14	Resources and Tools are defined.		
15	Schedules for performing software releases are defined.		
16	Software Configuration Management Plans are prepared to show configuration concepts.		
17	SCM tools are identified in the SCMP.		
18	Software Media will be marked and labeled.		
19	The Version Description Document (VDD) is prepared to document the configuration of the software products delivered to the customer.		
20	Processes or applicable SCM instructions are developed.		
21	SCM records and reports will be prepared and maintained as identified in the SCMP or applicable SCM instructions.		
22	Baseline release audits are conducted.		
23	Measurement data during baseline audit is collected.		

FURTHER READING

Keyes, J., *Software Configuration Management*, Boca Raton, FL: Auerbach Publications, 2004.

IEEE Computer Society, *Controlling Software Development*, 1996.

MIL STD-973, Configuration Management.

9

Software Supplier Audit

You must treat the software suppliers as an integral part of the team and manage the subcontracted software product development. Ensure that appropriate supplier software audits are planned and also monitor progress based on evidence of completion. The quality organization with the support of configuration management will conduct an audit of the supplier's quality, engineering, and management systems in support of a project as required by contract. The software supplier audit will be conducted through engineering review processes and procedures supporting the work defined by a purchase contract per the Statement of Work (SOW).

The audit will focus on major tasks, services, program phases, and events planned, as an example, for the following technical areas:

- Supplier Data Requirements Lists
- Configuration Management and Change Control
- Control of released documents
- Software design change processes
- Software Development
- Data Management
- Verification and Plans and Processes including Software Configuration Audit FCA
- Review of previous audit results and Corrective Actions

9.1 STARTING POINT FOR SOFTWARE SUPPLIER AUDITS

A company has decided to subcontract. There are key things to be aware of when working with a supplier, ultimately leading to acceptance of the software when delivered. There are key things to be aware of when working with a supplier, ultimately leading to acceptance of the software when delivered.

- Documented and agreed-upon supplier development and delivery process
- Statement in the Service Level Agreement that the supplier must be in compliance with the process (to contractually bind the supplier)
- Segregation of supplier work from company and program work
- Inclusion of acceptance test cases/reviews, SQA compliance, and security checks in the SOW
- Defined acceptance criteria, including audit and acceptance testing, of how the work will be validated
- Gate review at the point where the supplier has indicated the software is ready and has provided all the software deliverables

9.2 KEY REQUIREMENTS FOR SOFTWARE SUPPLIER AUDITS

When work takes place for a software company or military and aerospace program or project that has been subcontracted, the key requirements are as follows:

- Traceability and accountability, which provide the ability to resolve any contractual or security issues.
- Clearly established standards regarding where the work is performed and how the configuration software is used.
- A method to identify a specific release package; the label used must be recognized by others so release packages can be traced to the supplier.

9.3 GATE REVIEW FOR SOFTWARE SUPPLIER AUDITS

When the supplier has completed its development it is time to perform the gate review. The first element of the gate review is the audit. The audit ensures that the software code

- Is functional (code and peer reviews, executing agreed-upon test cases does not introduce regression problems, etc.)
- Contains no non-agreed-upon features (any functionality beyond the SOW)
- Is secure (no malicious code such as Malware, Spyware, viruses, etc.)
- Passes (preliminary) acceptance (functional within scope and secure)
- Passes SQA compliance as defined in the SOW

The supplier audit takes place in the supplier work area to maintain a clear division from the supplier and the company, program, or project work. Completion and passage of the supplier audit must be recorded. If it does not pass, the work should be rejected in writing to the supplier.

9.4 SOFTWARE TECHNICAL SUBCONTRACT MANAGEMENT

Software plans for managing subcontractors involved in the provision and development of software comprise methods used to monitor and coordinate subcontractor software development activities. The Integrated Product Team (IPT) approach is used to manage and coordinate with the subcontractors. With this approach, all team members are quickly informed of program events and are able to respond quickly to problems and concerns as they arise. Reviews and monitoring of the subcontractor's efforts will be performed by the appropriate IPT.

Metrics will be gathered on the subcontractor's efforts to monitor progress and risk areas. Table 9.1 shows the scope of development being undertaken by the major software development subcontractors. The requirements of this paragraph apply to the software subcontractors responsible for the software. Software programs require similar documentation (development and test plans, etc.) from its subcontractors and will monitor the plans and processes of their subcontractors to ensure these are in alignment with the many software approaches.

A Subcontract Management Plan (SMP) describes the detailed plans and processes used in the management of subcontracts.

TABLE 9.1

Software Subcontractor Scope of Development

Subcontractor	Scope of Development
Supplier A	Accomplish all software engineering tasks as necessary for the development, testing, and verification of the software associated with the following items: • Subsystem
Supplier B	Accomplish all software engineering tasks as necessary for the development, testing, and verification of the software associated with the following items: • Subsystem • Segment

9.5 SUBCONTRACTOR SOFTWARE CONFIGURATION MANAGEMENT AUDITS

The Subcontractor SCM audits I have performed are very helpful to software companies who have contracted to a supplier. Program QA always employs a software expert to perform the required SCM audits since QA is more interested in the hardware requirements. A program QA individual once said to me, "If you have made a supplier angry with you as you leave the facility, you have performed a very good supplier audit." I am not in total agreement, but at least you get the respect and appreciation of the subcontractor. Honestly, I have always left SCM audits with suppliers feeling good to know we have worked together to ensure that the software product met the requirements, design, and SCM disciplines for the company.

9.5.1 Subcontractor Configuration Identification

The subcontractor will prepare, maintain, and publish a released Configuration Identification data list identifying all released program data and software documentation related to each end item. Subcontractors also prepare, maintain, and publish a released item data list identifying all released program data and documentation not directly related to a program end item. The subcontractor release system identifies the entity with the authority to authorize the software release of a change/revision to an item of program-released data and documentation.

Subcontractor procedures require an authorized change/basic release, referencing to the change and Work Breakdown Structure (WBS) or program equivalent number in order to release an item of program data or documentation. The subcontractor procedures require that all engineering data related to a specific change be released. If not, how are additional releases against this change as scheduled, committed, or recorded tracked? Release records such as an Engineering Release Record (ERR) for each item released in the release system will be identified. Provide information on what software organization has responsibility for the distribution of software engineering data and documentation. The type of distribution is important (e.g., electronic, digital file part, hard copy) for delivery.

9.5.2 Subcontractor Computer Program Library

The subcontractor will describe how it will ensure that all software products, documentation, and engineering data approved for use in software development, integration, testing, production, or distribution/delivery will be placed under configuration control in the Subcontractor Computer Program Library (CPL). The supplier will identify how published procedures/process documentation describes how to

- Store, access, and update libraries.
- Protect their content against damage or loss.
- Support the level of control defined by the program.
- Demonstrate compliance with program contract/regulatory/program requirements.

9.6 MONITOR SUPPLIER SOFTWARE QUALITY

The following paragraphs describe the SQA activities used to support supplier management. These are applicable to software suppliers that are delivering products that have been developed or modified.

9.6.1 Monitor Supplier Quality Method

The purpose of the SQA method is to provide software programs with the process steps necessary to support software supplier management. This is applicable to suppliers that are delivering software that has been developed or modified for software programs, whether delivered separately or embedded as part of a system (firmware on hardware).

9.6.2 Supplier Selection Activities

Determine the quality requirements that are either imposed on software programs by the prime contract or regulatory agencies, or self-imposed and determine the extent of the development or modification effort expected from the candidate suppliers. Ensure that the request-for-proposal package addresses the determined quality requirements.

Support supplier selection by researching the current capabilities of the candidate supplier's quality system. Once a supplier has been selected, verify the appropriate contractual flow of quality requirements to the supplier, including any SQA-specific deliverables that are required. Establish an SQA Plan for Monitoring. Based on the supplier, coordinate with project personnel and plan the extent of monitoring the supplier and the internal supplier management group. Consider whether a Quality Management System Requirements for Supplier (QMS) survey is required, whether additional project-specific audits are required, whether formal/milestone reviews are to be attended, and whether the product qualifications or some portion thereof are to be witnessed. Based on the contract and other quality system requirements, determine the product acceptance criteria.

9.6.3 Provide Quality Plan

Coordinate with project personnel and plan the extent of monitoring to be performed on the supplier and the internal supplier management group. Consider whether a QMS survey is required, whether additional project-specific audits are required, whether formal/milestone reviews are to be attended, and whether the product qualifications or some portion thereof are to be witnessed. Based on the contract and other quality system requirements, determine the product acceptance criteria and document this SQA planning.

9.6.4 Quality Management Survey

When a QMS survey is required, look at the QMS or procurement representative and request that a survey be initiated. Annual surveys must also be planned throughout the supplier's product lifecycle.

9.6.5 Project-Specific Supplier Audits

According to the plan developed for monitoring the supplier quality system, perform project-specific audits of supplier management and at the supplier's facilities.

Coordinate audit schedules, audit notifications, and other related information with the supplier via the project supplier administrator. Perform and document audits per "Process Audits" in Chapter 7, with the exception that the audit results and findings will be coordinated with the supplier via the project supplier administrator.

9.6.6 Review Supplier-Delivered Products

An evaluation of software supplier products will be performed as required by contract, regulatory requirements, software development plans, and software quality engineering plans. Deliverable software products may include program plans, requirement specifications, design specifications, source code, executable code, verification/validation plans, test documents, and test reports.

9.6.7 Participate in Formal Milestone Reviews

According to the plan developed for monitoring the supplier quality system, participate in formal milestone reviews such as Preliminary Design Reviews (PDRs), Critical Design Reviews (CDRs), and Test Readiness Reviews (TRRs).

9.6.8 Support Supplier Qualification Test

Review supplier plans and processes associated with software qualification, whether deliverable or not, for conformance with contractual and functional requirements. Support the supplier formal test to ensure compliance to contractual requirements according to the plan developed for monitoring the supplier software quality system.

9.6.9 Acceptance of Software Supplier Reviews

Ensure the review of supplier data, such as version description documentation and software test records, in accordance with the program software development plans. Acceptance is granted, in accordance with the program software development plans, upon completion of all related contract delivery requirements and satisfaction of the planned product acceptance criteria. An example of a Software Supplier/Subcontractor Checklist is provided in Table 9.2.

TABLE 9.2

Software Supplier/Subcontractor Checklist

Item	Description	Y/N/NA	Notes/Comments
1	The Software Supplier SDRL is defined.		
2	The Supplier Software Development Plan (SDP), Software Configuration Management Plan (SCMP), Software Quality Assurance Plan (SQAP), and any additional supplier planning documents are complete and released.		
3	Software residing in firmware is identified and developed in accordance with the supplier's SDP.		
4	Prepared comments and action items on supplier planning documents have been reviewed.		
5	Metrics data required by the supplier's contract is available and reviewed.		
6	Schedule, quality, process, and technical/functionality issues are identified based on review of supplier metrics data.		
7	Supplier corrective action plans identify issues and are reviewed to be complete and support closure.		
8	Supplier progress on corrective action plans shall be monitored and applicable to ensure that action items are addressed or mitigated.		
9	Software-related action items for supplier program management reviews and summary program/project reviews are resolved or tracked to closure.		
10	Milestone reviews, such as Preliminary Design Review (PDR) supplier criteria, are defined.		
11	The Software Development File (SDF) or equivalent software development repository, as applicable, is reviewed.		
12	Test Readiness Review (TRR) coverage and regression test/coding analysis is verified per the contract.		

Continued

TABLE 9.2 (*Continued*)

Software Supplier/Subcontractor Checklist

Item	Description	Y/N/NA	Notes/Comments
13	Supplier software qualification testing has been performed to the approved Software Test/Verification Procedures and observed/monitored in support of Quality Assurance.		
14	The FCA/PCA or equivalent (milestone review) will be conducted and performed utilizing defined and documented procedures and standards.		
15	Supplier system requirements have been reviewed.		
16	Supplier processes are reviewed.		
17	Supplier software process issues and action items are tracked to closure.		

FURTHER READING

Department of the Air Force Software Technology Support Center, Guidelines for Successful Acquisition and Management of Software Intensive Systems, Weapon Systems,
Command and Control Systems, Management Information Systems, Version 1.1, Volume 2, Appendix M, A Detailed Comparison of ISO 9001 and the Capability Maturity Model (CMM), February 1995.

10

Software Engineering Reviews

When software organizations and military and aerospace programs change their ways of conducting business, the change cannot be trouble-free. The addition of new ideas can increase stress, while adding monthly software engineering reviews to committed programs or project schedules. This does not mean that making a change is a bad idea. Monthly reviews serve as a method to determine and provide verification of the products generated by the program or project before conducting and performing formal audits (i.e., First Article Inspections [FAIs], FCAs, and PCAs). Verification is a key process to ensure correctness and consistency with respect to the tasks performed and to ensure that documented plans, processes, and procedures are compliant.

10.1 RESOURCES

Senior Management, who lead software industries and military and aerospace programs, must ensure that adequate resources and funding are provided to support these incremental reviews. Assignments are determined by responsible organizations, and must be confirmed and understood. This includes adequate staffing levels on programs in addition to overhead for military and aerospace or business infrastructure. Resources include the required tools to perform the engineering reviews. Funding can be an issue. I have seen funding and budgets cancelled when work orders are not in place. The right personnel and funding, and effective tools help software teams and businesses to be successful.

An example of funding issues can be seen in sport teams (i.e., professional and college) that do not have the right personnel in place to be successful in the business world. Professional owners or funded colleges must look at these important points before venturing into the market of competitive competition. I have attended Major League Baseball (MLB) games and noticed that good pitching always stops good hitting and vice versa. Have you ever heard that statement before? A great defense in the National Football League (NFL) will stop a great offense and vice versa. If you win, more fans come out to the games. Funding and success will continue year after year if the necessary tools are used and implemented in order to be successful. Enough about sports; I could go on and on about professional and college sports. I love watching sports and talking with my friends about the execution of my favorite teams. Oh, boy! Get it, execution? Having the right resources—personnel, funding, and tools—in place will make a program or business succeed.

10.2 SOFTWARE ENGINEERING REVIEW STEPS

This chapter provides the necessary steps and mitigates important software problems or concerns to be eliminated when it comes time to perform internal software engineering audits, which lead to formal audits. I will address the Integrated Product Team (IPT) throughout this chapter.

10.3 PURPOSE

The purpose of the Software Engineering Review method is to provide software companies and military or aerospace programs the process steps necessary to perform incremental reviews during the software lifecycle. Required actions, including informal or incremental software engineering and supplier or subcontractor reviews, are described in this chapter.

Software Quality Assurance (SQA) is the key organization to establishing a quality foundation upon which a program's or project's performance is based. The objectives of SQA monthly reviews and evaluations ensure that software tools, plans, processes, and procedures support the goal of producing products and providing services that meet predefined quality criteria inside the program. The next section identifies the fundamental objectives of SQA.

10.3.1 Software Quality Assurance

A fundamental objective of SQA is the continuous improvement in the quality of products and processes. Process improvements are achieved by defining, documenting, measuring, analyzing, auditing, and improving the development processes in order to reduce error rates and flow time. I enjoy talking about and presenting SQA to the software world. Software organizations worldwide want to know what SQA is all about and how to implement quality into their software programs or projects. At this time I am working with international customers who play a key role in software development activities. The SQA role is as follows:

- Establish and maintain SQA budgeting and estimating data.
- Establish communication channels with management.
- Integrate quality into the IPTs and encourage employee involvement in identifying quality issues.
- Plan for monthly review status delivery to management.
- Coordinate activities with the IPTs and management.
- Team with other functional and project groups and suppliers.
- Participate in the preparation of the plans, standards, and procedures.

10.3.2 Integrated Product Teams

The Integrated Product Team (IPT) is a multidisciplined team that may include supplier membership and has assigned responsibility for a specific product or subsystem. The IPT is accountable and is composed of members from the appropriate functional disciplines (e.g., Engineering, Supplier Management, Product Support, etc.) necessary to accomplish day-to-day activities.

The IPT is assigned authority and allocated a budget to perform activities derived from the Statement of Work (SOW), and is defined in team charters/execution plans.

10.4 PLANNING

At the start of each review period, the SQA focus group supporting the IPT will provide a plan to the IPTs by identifying plans, processes, procedures, and IPT interviews to be performed. The identified documentation and personnel are the criteria to support the reviews. During the review planning, SQA completes a monthly schedule. The planning includes the following:

- Plans, processes, and procedures to be reviewed
- Obtaining the appropriate schedules, plans, and procedures for the IPT
- Determining which processes should be evaluated during the review period
- Scheduled interviews with IPT personal

The planning requires report and completion dates for the review period. SQA will record the performed activities each month, the total number of reviews, and the interviews and provide a copy of the completed review plan to the IPT manager.

The plan will be updated, as needed, to reflect any changes in the IPT development schedule or activities. At the time the updates are made to the plan, the report date will also be updated. Updates to the review plan will be provided to the IPT manager.

The IPT or Supplier Management (SM) that manages suppliers or subcontractors will provide the review plans to SQA for review and evaluation.

10.4.1 Software Development Plan

Plans, procedures, and objectives for software development are documented in a required Software Development Plan (SDP). Details of components, functionality, and their associated schedules are contained in the SDP.

This plan will be updated, as required, to ensure that it remains consistent with processes, development methods, tools, procedures, and standards as they evolve. The identification of standards that result from prototyping activities will be reflected in updates to the appropriate software build plans and will be updated to reflect corrective actions and re-planning activities.

Each major software development activity will be planned in accordance with the steps outlined in the SDPs, and will include the following:

- Definition of entry and exit criteria for the activity
- Review and assessment of the product and task requirements
- Definition or update of the process for the each activity
- Development or update of the estimating process

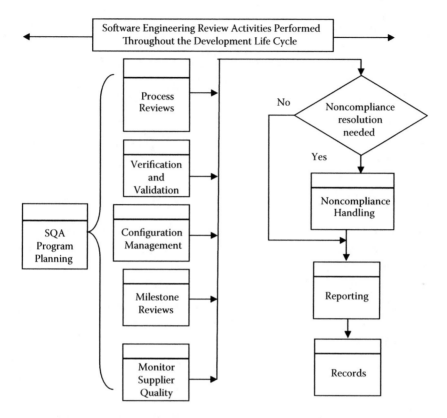

FIGURE 10.1
Software Quality Assurance review process.

- Development of the initial cost and schedule estimates and risks
- Preparation of the detailed implementation plans

The SDP will comply with the organization's standard software processes. SQA personnel conduct and perform the software engineering reviews based on the SDP. An overview of the SQA engineering review process is shown in Figure 10.1.

10.4.2 Review Plans

The Review Plans' incremental reviews are scheduled and performed by the SQA focus group for each IPT. These reviews form the basis to ensure compliance that development activities performed are in accordance with plans and procedures, and are in line with requirements. There are six steps in completing the incremental software engineering review:

1. Planning
2. Performing
3. Recording
4. Reporting
5. Corrective Action
6. Process Improvement

10.5 PERFORMING

SQA, using criteria derived from documentation plus any other plans, processes, procedures, and interviews, are documented monthly. The purpose of the software engineering reviews is to ensure that development activities are completed as planned and are compliant. SQA will evaluate how the suppliers or subcontractors are performing as well. SQA evaluates the supplier's or subcontractor's development processes. Performing reviews for the IPTs includes

- Reviewing the plans, processes, and procedures to determine and select appropriate evaluation criteria
- Reviewing and analyzing the results of previous reviews
- Performing the review using the selected criteria
- Interviews of IPT members
- SQA making an assessment as to whether the implemented process is compliant or noncompliant or recurring noncompliant
- SQA identifying an opportunity for improvement
- SQA rescheduling the review if it is unable to perform the review as planned

10.6 DOCUMENT CONTROL PROVISIONS

SQA Management will review the reviews for completeness and accuracy, as related to existing requirements, on a continuous basis. Document control is exercised via document number, revision letter, and date. Programs and projects that support Software Engineering Reviews will participate in defined initiatives, developing products, and intermediate products that have been identified as a Configuration Item (CI) or a deliverable software item.

The primary goal of the engineering review is to mitigate the initiation of excessive and unresolved action items before an informal audit. The implementation of scheduled engineering reviews during the software lifecycle will benefit the contractor. The engineering review will ensure, in a planned schedule, a smooth transition to software acceptance, and delivery of quality products to the customer. A qualified team under the direction of SQA will perform the engineering reviews internally in the IPT organization. The review team will provide assurance that required or contractual data is reviewed and released, and that interviews and checklists are complete and action items are closed and verified before the scheduled informal audit is performed. The software engineering review checklist required is defined in Table 10.1.

10.7 RECORDING

The results of internal and monthly software engineering reviews are recorded and maintained in the SQA database. Recording the status of each of the evaluated reviews includes the following:

- Compliant (C) – The implemented software development activities have been successfully accomplished in accordance with reviewed documentation, associated plans, and procedures.
- Noncompliant (N) – The implemented software development activities differ from the process defined in the documentation, associated plans, interviews, and procedures.
- Opportunities for Improvement (O) – The implemented software development activities have been successfully performed in accordance with the documentation, associated plans, and procedures. An opportunity to improve the process has been identified.
- Planned Event (P) – Planned Software Engineering Review to be performed.

SQA Records reflect the results of planned/scheduled software engineering reviews that have occurred and will be recorded in the SQA database per Table 10.2.

TABLE 10.1

Software Engineering Review Checklist

Item	Description
1	The scope of the review is clearly defined to cover one or more software end items.
2	Presentation material for the review is clear and sufficient in detail and consistent with the scope of the Software Engineering Reviews.
3	Review of the data to be available, including the following: • Software Requirements Definition • Software Interface Definition • Software Design Definition • Version Description Document • Code and Unit Test Results • Software Integration Test Results • Software Test Plans • Software Test Description (including software test cases and software test procedures) • Software Test Results • System Test Results • Software Requirements Traceability • Inspection and Demonstration Results • Analysis Results
4	Action items for software from previous reviews have been completed and issues are resolved.
5	A current listing of all deviations/waivers against the software end item, either requested of or approved by the procuring activity, is available for review.
6	The "as built" configuration of the software end item that underwent software testing is identified in software release documentation and is available for software review.
7	The status of the software testing environment, hardware, and software used to test the software end item used for software testing are identified and available for review.
8	The software requirements definition for the software end item is complete.
9	The implementation status of changes against the software end item is identified and the changes are available for review.
10	Traceability between the software requirement, verification method, and verification evidence (i.e., testing, demonstration, inspection, or analysis data) to show compliance to the software requirement is complete and available for the review.
11	The software testing for the software end item(s) is complete.
12	All problem reports and issues identified during software testing are resolved.
13	The software testing results for the software end item are complete and available for review.
14	The "as run" software testing procedure is complete and available for review.
15	Evidence used from software code and unit testing and software integration testing to verify that a software requirement is under configuration control is complete and is available for review.

TABLE 10.2

SQA Database

Months in a Year	Jan.	Feb.	Mar.	Apr.	May	Jun.	Jul.	Aug.	Sept.	Oct.	Nov.	Dec.
Process Title												
Planned Software Engineering Reviews	1	1	1	1	1	1	1	1	1	1	1	1
Product Evaluation	C								P			
Software Configuration Management		N								P		
SW Architecture			C								P	
SW Engineering Environment				C								P
SW Design					C							
Defect Management						P						
SW Quality Assurance							P					
Process Deployment								P				
C = Compliant	1	0	1	1	1	0	0	0	0	0	0	0
O = Opportunity for Improvement	0	0	0	0	0	0	0	0	0	0	0	0
N = Noncompliant	0	1	0	0	0	0	0	0	0	0	0	0
P = Planned Process Evaluation	0	0	0	0	1	1	1	1	1	1	1	1

10.8 REPORTING

The reporting of software engineering review results produce weekly or monthly reports, which are provided to the IPT manager, SQA Manager, and QA management.

Reporting includes the following:

- The weekly report or a monthly report.
- The weekly/monthly IPT report provides a discussion of any issues, noncompliances, opportunities for improvement, etc., that were identified during the monthly engineering review and the status of all open items requiring a Corrective Action (CA) response.

10.8.1 Reporting and Coordination

The reporting and coordination of software engineering reviews instills quality and maintains process/product requirement compliance and oversight to promote process improvements, and advocates a quality culture that supports a commitment to technical integrity. Ensuring that internal software engineering reviews are performed will drive the growth of individuals, programs, and projects. Personal and professional software development will be focused on disciplined execution.

10.8.2 Corrective Action/Process Improvement

Noncompliant processes are documented as a Corrective Action (CA)/ Process Improvement. Process improvements will be recommended by SQA and administered by IPT and QA management.

10.9 SUMMARY

Military and aerospace programs need to implement software engineering reviews during the software lifecycle prior to informal and formal audits. Make sure these reviews are planned and scheduled incrementally for each software subsystem, segment, or system during the software lifecycle. The role of senior management will determine what organization will conduct and perform the reviews during the software lifecycle. This concept will shake up defense and space personnel, but in the long run it will benefit contractual deliveries to software companies and future military and aerospace programs. SQA needs to be the organization to conduct and perform the incremental and monthly software engineering reviews. Did you hear me say SQA?

I will not provide many references for this subject, due to the experience of conducting and performing Software Engineering Reviews. We need to have these critical reviews before proceeding to Informal Engineering and Formal Audits. I have more to say about this subject, but I would like to present more information at future software and technology conferences. That is where we can discuss the critical issues or roadblocks and come up with solutions as an audience of interested groups or companies.

Let us now move on to the next chapter and discuss software engineering audits and their importance to software companies and military and aerospace programs or projects.

FURTHER READING

Humphrey, W.S., *Introduction to the Team Software Process*. TSP:SM, Reading, MA: Addison-Wesley Longman, 2000.

Morris, R.A., *PMP, The Everything Project Management Handbook*, 2nd ed., Avon, MA: Adamsmedia, 2008.

Wellins, R.S., Schaaf, D., and Shomo, K.H. *Succeeding with Teams*, Minneapolis, MN: Lakewood Publications, 1994.

IEEE STD 028-1988, IEEE Standard for Software Reviews and Audits.

11

Software Engineering Audits

This chapter describes the tasks, responsibilities, processes, and procedure methods for accomplishing an internal software engineering audit. The audit will verify that the actual performance of the software complies with applicable system, segment, and subsystem requirements. The software engineering audit methods provide team members an understanding of quality in supporting performed product evaluations inside the program or project.

11.1 AUDIT OVERVIEW

An audit team (i.e., Systems Engineering, Engineering, Quality, Configuration Management, IPTs, etc.) will periodically perform internal or incremental reviews of the program or project processes, procedures, and products. This audit serves as a method to determine how effective and efficient the internal processes and procedures fulfill the requirements defined in applicable plans.

Series of meetings with team members combined with applicable discussions ensure that the approach is consistent. For software industries and military and aerospace programs or projects, the software engineering audit methods discussed in this chapter ensure compliance and quality in developed products before starting contractual formal audits. Let me repeat! The software engineering audit is performed before conducting and performing a formal audit (i.e., FAI, FCA, and the PCA). You could call this audit a readiness review, but do not start formal audits until all addressed action items from the software engineering audit are complete and closed. The segment and subsystem software engineering audits performed internally require available budget, engineering due diligence, and a documented approach in the audit plan. Internal documents, extracts from plans, processes, and procedures will be the required documentation. Interviews with IPT members will also be required.

11.2 AUDIT SCOPE

The scope of the software engineering audit is to audit and confirm that required information is complete and correct, and that configuration changes have been sufficiently addressed. In order to perform a software engineering audit, the following prerequisites must be met:

- Audit Preparation
- Audit Plan
- Audit Scope
- Collect Data Package
- Conduct Audit
- Audit Exit Criteria
- Audit Action Items

11.3 AUDIT PREPARATION

The software audit is prepared and held to verify that the performance of the deliverable software complies with the requirements and applicable design descriptions. The test procedures and results are reviewed to verify that the software performs as required by its Functional/Allocated Configuration Identification. The audit is conducted after a major change or a significant number of minor changes have occurred, or before the establishment of a software product baseline.

The software audit is also an informal audit to verify that the Configuration Item(s) (CI) or computer software actual performance complies with the hardware or software performance and interface requirements. The test and analysis data is reviewed to verify that these allocated requirements are met by the hardware or software configuration item(s). In addition, incremental completed operations and support documents will be reviewed.

11.4 AUDIT SCHEDULE

The audit schedule will be defined in the software plans. This audit will be conducted and performed in accordance with internal program or project plans. Software Quality Assurance (SQA) or Software Configuration Management (SCM) will conduct the software engineering audit. In my experience with military and aerospace software programs, there has been confusion about who will conduct the internal audit. Is it quality, configuration management, systems engineering, or the IPT leads? It all depends on how plans, program instructions, and software standards define the software program or project objectives. This is the reason I am writing this book, to help software industries and military and aerospace programs to develop a plan for who will conduct the informal audit, how these audits are to performed, and provide implementation criteria.

11.5 AUDIT PLAN

The Audit Plan describes the scope of each planned, schedule, and preparation for the audit, providing identification of the item(s) to be audited, associated contract requirements, and drawings or documentation to be available. Reviews are performed per the internal audit plan. These audits are performed in accordance with the software processes documented and for the following audit activities:

- Perform internal process audits to ensure compliance of defined software processes.
- Perform internal audits to ensure compliance.
- Conduct interviews with team or IPT members and management.
- Support internal audit exit criteria.

The audit team will verify that the product design has been accurately documented and will document any findings and submit them for corrective action to the IPT.

11.6 AUDIT COMMUNICATION

Prior to the internal engineering audit, the audit team will communicate with affected team members to provide a basis for understanding and agreement. An opening meeting (informal or formal) will be conducted to describe the audit process, expectations, and establish appropriate communication channels. During the internal audit, the auditor(s) will periodically communicate progress and preliminary issues/concerns, including potential nonconformity and/or effectiveness issues with affected IPT members. Any concerns affecting product quality will be reported immediately to responsible management and appropriately addressed.

11.7 AUDIT ENTRY CRITERIA

Audit entry criteria provide coordination and administrative tasks to be established. Reserved conference rooms are set up for the audit team to meet and discuss findings or issues pertaining to the software engineering audit. An Engineering Data Package is compiled (plans, procedures, specifications, verification artifacts, etc.). The Engineering Data Package (EDP) content is confirmed to be ready when a self-audit of the contents is 100% complete.

11.7.1 Collect Data Package

Integrated Product Teams (IPTs) and other affected organizations will collect and provide a required Engineering Data Package to support the audit team and ensure fulfillment of the audit requirements. Also required are formal test plans and reports to be submitted with other verification data. Some of the most commonly used documents that will be made available to support the internal audits are as follows:

- Copies for the System Requirements Review (SRR) and System Design Review (SDR) documentation.
- Preliminary Design Review (PDR), Critical Design Review (CDR), Test Readiness Review (TRR) documentation.
- Interface Control Documents (ICDs).
- Approved operating plans and test procedures including detailed verification requirements.
- Completed qualification test, verification inspection, and analysis reports.
- Completed compliance matrix showing traceability requirements.
- Traceability Reports from the specifications down to the hardware and software.
- The software audit briefing, including an agenda.

Note: Other supporting data should be quickly accessible upon request.

Often when it is time for a software audit to be conducted, there is panic and personnel are not sure what an engineering data package is. Working in software programs, it is critical for the IPTs to know what is needed. It is important to have an understanding of what is required, and I say this again and again, know what the audit requirements represent. Selection of the right personnel is important; make sure they have a strong commitment to participation in the audit. Some software organizations or IPTs have no interest in participating in internal software engineering audits. That is where senior management should step in and make sure that audit schedules are in place and ensure that IPTs are able to support them.

The audit team will include everyone directly involved in the successful completion of the software audit on any given CI or software.

11.8 CONDUCT AND PERFORM AUDIT

The audit team, using criteria derived from the collected data package, will conduct and perform the internal software audit. The purpose of the audit is to ensure that software development activities are being worked as planned and are compliant with approved plans and procedures. Performing the audit involves the following:

- Reviewing the plans and procedures to determine and select appropriate audit criteria
- Reviewing and analyzing the results of previous audits
- Performing the audit using the selected evaluation criteria
- Interviewing team members or IPT personnel

In performing the audit, auditors will make an assessment as to whether implemented processes are compliant, noncompliant, or recurring noncompliant. During the audit, the auditor may identify an issue or opportunity for improvement, or identify the process as a trusted process. The audit findings are recorded and the status of each will be indentified. The auditors will provide totals for each finding and evaluate and document evidence for the audit. Table 11.1 provides an example of a software engineering audit checklist. There are more audit questions that can be developed and used in a software engineering audit.

During the internal audit, the auditor(s) should periodically communicate progress and preliminary issues/concerns, including potential nonconformity and issues with affected teams, IPTs, or organizations. Any concerns will be reported to senior management and be appropriately addressed.

TABLE 11.1

Software Engineering Audit Checklist

Item	Description
1	Audit scope and criteria will be defined from previous audit history, organization procedures, development plans, or other documented requirements.
2	Audits shall be scheduled based on program or project schedules and activities.
3	Results will be recorded, including any noncompliances or observations.
4	The audit report will contain, at a minimum, the area audited, scope and purpose of the audit, completed checklists, audit criteria, participants, results, noncompliances/observations, and any lessons learned for future improvement.
5	Audit reports will be approved by SCM management or its designee and distributed to relevant stakeholders.
6	Any noncompliance will be addressed.
7	Actual measurement data generated during the conduct of the SCM audit will be collected.
8	Applicable work products will be submitted for control in accordance with software development plans.

11.8.1 Audit Action Items

The audit team will document and implement action items to resolve questionable or problem areas for preparation of the audit report. The audit team will prepare a summary report identifying action items initiated during the internal audit and identify responsibilities for resolution. The summary will be provided to all affected team members, IPT leads, and support organizations to conduct follow-up actions to ensure that all post-audit action items are closed and report the status to senior management. The audit lead will require support from the audit team in developing recommended solutions to actions/issues along with requests for any waivers, deviations, or corrective actions required in support of the audit. The audit team will prepare and sign off on the audit sheets, and prepare and submit the software audit minutes showing closure of the action items.

11.9 AUDIT REPORT

A closing meeting (informal or formal) may be conducted with senior management and affected teams to communicate the internal audit results and provide clear direction on corrective action expectations.

The audit team leader will present the audit report using the items listed in the assessment report and the audit checklist at a minimum, stating its conclusions on the conformance and effectiveness of quality standard requirements. The audit team will leave copies of all information pertaining to the internal audit results (including checklists, findings, supporting documents, and other correspondence) for the purpose of sharing this information with the organization or customers. The results of the internal software audit report are recorded in the audit record and identify the status of each evaluated process as one of the following:

- Compliant
- Issue
- Noncompliant
- Opportunity for Improvement
- Recurring Noncompliant
- Rescheduled Evaluation

11.10 AUDIT EXIT CRITERIA

The audit team will review the proposed nonconformities and other applicable information collected during the internal software audit in support of satisfying audit objectives. When a process deviation identified as noncompliant has an adverse impact upon the process being completed as defined in the documentation (or associated plans and procedures), the software audit team will initiate a Corrective Action (CA). The process owner, along with the IPT manager and others identified as being impacted, are notified. A Corrective Action Plan (CAP) is requested and is due on a specified date. The noncompliant process is then monitored until compliance has been verified and documented. Failure to respond or complete the CAP will result in the problem being elevated to senior management.

When an opportunity for improvement has been identified, the auditors will document the process as an opportunity for improvement. The process owner, along with the IPT manager, is notified. If the process owner accepts the opportunity, the process will then be monitored for implementation of the approved improvement as required; otherwise, the opportunity is closed. The audit team will verify that the improvement has been implemented and document the implementation of the improvement. Agreement will be reached when conclusions, nonconformities, and any other suggested recommendations including positive observations are addressed. The software audit team leader will sign off on the exit criteria and meeting minutes and then the audit action items are closed.

11.11 AUDIT RECORD RETENTION

Internal audit records are maintained and retained for retrieval by organization team members or IPTs who have a need to know or who will use the audit records to analyze the quality process and make improvements as required. The audit records should be accessible to all audit team members to establish, maintain, and communicate awareness and other necessary requirements. Management of records and information includes the following:

- Identifying the records created or received and maintained in a records control index
- Providing current policies and procedures
- Maintaining records throughout the software lifecycle
- Keeping records in an approved off-site records storage center
- Adequately controlling and protecting records to prevent their loss, damage, or unauthorized use
- Utilizing a process for collecting records as created to ensure the records can be retrieved
- Retaining records only for the identified and specified period of time

Senior management and IPT management will manage records in their area of responsibility and maintain the audit records. They will protect records in their area of responsibility and ensure that they are available for review to support the next audit team. It is important to ensure that records containing classified or controlled documentation are protected. The internal audit is closed when all corrective actions or action items have been implemented and follow-up is completed. The following documents will be retained as records of supporting evidence of the internal audit, the results, and conclusions of the audit and associated corrective actions:

- Audit Schedule
- Audit Plan
- Audit Report
- Corrective Action (CA)
- Evidence of CA closure
- Evidence of CAP Closure

11.12 SUMMARY

Military and aerospace programs working towards a scheduled First Article Inspection (FAI), Functional Configuration Audit (FCA), and Physical Configuration Audit (PCA) with qualified auditors analyze issues identified per required audit processes. The root cause can be analyzed and identified, and corrective actions are undertaken to prevent recurrence. Conducting and performing informal audits prior to formal audits in military and aerospace programs is the main purpose of this chapter. I have described in detail the importance of scheduled software engineering audits early in a program and the benefits of having formal audits go off without a hitch instead of being performed over and over again before acceptance by the customer, which is a major cost to the contractor and customer.

Along with interested software managers and software engineers, I am working in a Boeing assigned team called a High Performance Work Team (HPWT). This team is working towards developing an effective and constructive process for performing formal audits. Included as an example is the Physical Configuration Audit (PCA) for software to be evaluated by our team. Also, provide an audit team checklist required to meet contractual and program requirements for formal software audits. In the HPWT there are many suggestions and questions the team is trying to answer. Should configuration management and quality people for hardware organizations be assigned a leader for all the formal audits for software? These organizations should conduct and perform the required and contractual formal audits, but for software, leave us (software geeks) alone!

I will give you my opinion of what the roles should be for conducting and performing major and contractual formal audits, both "Hardware and Software."

Boyd L. Summers

Is it a requirement of Software Quality Assurance (SQA), dedicated to military or aerospace programs, to lead formal audits for software? It comes down to this: what is the role of quality assurance/engineering in conducting and performing formal audits related to software? It is time for SQA to step up and take responsibility for preparing to conduct and perform formal audits (i.e., FAI, FCA, and PCA) pertaining to software.

This activity or function is a sore subject with me and with other software industries and military and aerospace program personnel. But first, we need to understand what is required to conduct and prepare for effective formal audits.

A good way of understanding software engineering audits is through an example of a written and documented Software Development Plan (SDP). An SDP written and implemented in military and aerospace software programs starts the process for the software development life-cycle. Internal software engineering review audits are performed per the SDP. In the development of this plan, software guidelines for effective software development are defined. Just a note of interest, commercial and software companies or industries can learn from this chapter how to start implementing software engineering audits. Performing and under-standing engineering audits can lead to success for software businesses/industries and not just for military or aerospace software programs. As stated in the previous chapter, I will not provide many references to this subject, due to the necessary experience of conducting and performing Software Engineering Audits. We need to perform these critical engi-neering audits before proceeding to Formal Audits. I have more to say about this subject, but I would like to present more information in future software and technology conferences, where we can discuss the critical issues or roadblocks and come up with solutions as an audience of inter-ested groups or companies.

Let us now move on to the next chapter and discuss Formal Audits and their importance to software companies and military and aerospace pro-grams or projects.

FURTHER READING

Keyes, J., *Software Configuration Management*, Boca Raton, FL: Auerbach Publications, 2004.
MIL STD-973, Military Standard, Configuration Management.

12

Formal Reviews and Audits

Formal Reviews provide programs or projects the necessary steps to support applicable and scheduled milestones. They also provide a forum for design evaluation with experienced participants from other organizations during design development. The purpose of these meetings is to review the status of preparation for the formal design reviews and ensure that the design review objectives are accomplished. Meeting representatives from the responsible project groups and technical integration, and personnel from any other affected groups will attend. Plans are discussed and developed for collection, sharing, and closure of action items from the review(s). If customers are attending, involve them in developing appropriate plans or actions discussed in the Formal Review. In Formal Reviews, presentations and discussions are staged, and meeting attendance and action items are recorded. Initiate and coordinate follow-up for each action item assigned during the Formal Review. Upon resolution of each review action item, provide appropriate information to the presenter.

In order to maximize quality, the following agenda items for Formal Reviews are provided:

- Introduction
- Scope
- Actions
- Agreement
- Meeting Closure

12.1 SYSTEM REQUIREMENTS REVIEW

The functional baseline is comprised of the system and segment specifications. It is established upon the baseline during the System Requirements Review (SRR). The required documents are placed under high-level change control. The functional baseline is augmented at subsequent lifecycle reviews and is maintained and updated under configuration management control.

12.2 SYSTEM DESIGN REVIEW

At the System Design Review (SDR), change control is implemented to the functional baseline. The functional baseline describes the system's functional, physical, and interface requirements, which consist of architectural, functional, and performance decomposition. Physical and functional lines are determined by segment requirements.

12.3 SOFTWARE SPECIFICATION REVIEW

The purpose of the Software Specification Review (SSR) is to demonstrate to the customer the adequacy of the software and interface requirements to proceed into the design phase.

12.4 PRELIMINARY DESIGN REVIEW

The purpose of the Preliminary Design Review (PDR) is to determine if the top-level design of the software is mature and complete enough to advance to the detailed design phase. During the Preliminary Design phase, the system-level architecture, interfaces, and design are developed. A PDR is held and approval is obtained before proceeding with the detailed design phase. Activities include the following:

- Establish and maintain Software Development Library (SDL).
- Establish corrective action process for developmental configuration.

- Attend PDR.
- Exercise configuration control of developmental configuration products.

12.5 CRITICAL DESIGN REVIEW

The purpose of the Critical Design Review (CDR) is to determine if the detailed design of the software is correct, consistent, and complete enough for development to continue with coding and informal testing. This technical review is held to provide a detailed basis for verifying design integrity and compatibility with CSCI requirements and assessment of formal test preparation.

12.6 JOINT TECHNICAL REVIEWS

The Joint Technical Reviews, both formal and informal, are used throughout the program lifecycle to review and assess technical product quality and the development progress. Table 12.1 summarizes the types of technical

TABLE 12.1

Joint Technical Reviews

Description	Commercial Standards	Software Standards
In-process joint technical and management reviews of interim work products focus on program/project's development status, approach, and risks: • System Requirements • Software Development • Documentation Status • Software Development Folders • Software Coding and Testing System Integration and Testing	Hierarchical software designs are not imposed. Accommodates software design methodologies for analysis and design. Ordering of activities is dependent on lifecycle model used for incremental and evolutionary standards.	Software decomposed into "software units," which may or may not be related in a hierarchical manner. Ordering of activities is dependent on the software lifecycle model used to describe incremental and software standards. Addition to that area is the Waterfall software lifecycle influence.

reviews performed, their purpose, customer roles, and when these reviews are performed. Technical excellence in software products and services is critical to the success of software programs. Productivity improvements must be combined with first-time quality and a focus on technical excellence, while still meeting customer expectations in all programmatic dimensions. Table 13.6 defines which Joint Technical Reviews are performed and provides a comparison to commercial standards. A Formal Review results in a report identifying the material and the technical reviewers to ensure that the software product is in compliance and acceptable. The report is a summary of the defects found and concerns presented. Members of the Joint Technical Review team share in the quality of the review, but the software designer is responsible for the quality of the product.

12.7 INTERIM AND FINAL DESIGN REVIEWS

Software designs and development for newly developed functionality will be reviewed at the IDR and the FDR. Charts and materials are reviewed will be presented by the appropriate software development organizations The IDR and FDR focus on the unique requirements and incorporation of these capabilities into the design baselines, including the areas of requirements allocation, design architecture, environment, risk management, software development, safety, and product support.

12.7.1 Final Design Review Objectives

The objective of the FDR is to establish a product baseline that is compatible with contract requirements. An FDR completes the final design and forms for production baselines. The FDR presents a brief description of the previously developed system baseline to establish a framework that will allow a focused presentation on the final design of the system. The FDR also defines the detailed design and plans. At the end of the detailed design activity, confirmation will identify the system approach down to the subsystem level with each segment and demonstrate that the detailed design of the system satisfies system requirements. An FDR allows customers to assess the final design for the portions of a system.

Software documentation promotes software quality. One of the major goals of software engineering is to produce the best possible software along with the best possible supporting documentation. Software documentation can be either electronic or paper. Data Management is the organization in a military or aerospace program that will manage the software documentation for planned releases, customer requests, and storage.

12.9 REQUIRED SOFTWARE DOCUMENTATION

Software documentation promotes software quality. One of the major goals of software engineering is to produce the best possible software along with the best possible supporting documentation. Software documentation can be either electronic or paper. Data Management is the organization in a military or aerospace program that manages the software documentation for planned software releases, customer requests, and storage.

12.9.1 Software Documentation Status

There are software programs that follow no or few policies and procedures, and loosely follow standards. Good software development is standards based—let me repeat, standards based—and thus software documentation will be documented and become effective per the software development lifecycle.

12.9.2 Software Requirements Specification

The Software Requirements Specification (SRS) specifies the requirements for the software and the methods to be used to ensure that each requirement has been met. Requirements pertaining to the software external interfaces may be presented in the SRS or in one or more Interface Requirements Specification (IRS). The SRS, possibly supplemented by the IRS, is used as the basis for design and qualification testing of software.

Functional requirements are documented in a software requirements specification, which describes as fully as necessary the expected behavior of a software system. The SRS could be a document, database, or spreadsheet that contains the software requirements information stored in a requirements management tool.

12.9.3 Interface Design Description

The Interface Design Description (IDD) describes the interface character-istics of one or more systems, subsystems, Hardware Configuration Items (HWCIs), computer software, manual operations, or other system com-ponents. An IDD may describe any number of interfaces. The IDD can be used to supplement the System/Subsystem Design Description (SSDD), Software Design Description (SDD), and Database Design Description (DBDD). The IDD and its companion Interface Requirements Specification (IRS) serve to communicate and control interface design decisions. Also, the IDD may reference the IRS to avoid repeating information.

12.9.4 System/Subsystem Design Description

A System/Subsystem Design Description (SSDD) describes design deci-sions regarding subsystem behavior and allocation to subsystem/CI-level physical entities. Herein, the term "system" may be interpreted to mean "subsystem" as applicable, and an example of this checklist is based on a tailored MIL-STD-498 data item description DI-IPSC-81432A.

Checklist for SSDD

Review	Y	N	N/A
Does the paragraph contain a full identification of the System and the software to which this SSDD applies?			
Does this paragraph describe the purpose and general nature of the system to which the SSDD applies?			
Does this section list the number, title, revision, and date of all documents references in the SSDD?			
Does this section list the number, title, revision, and date of all documents references in the SSDD?			
This section addresses the baseline system and, if required, the proposed changes to the baseline system.			

12.9.5 Database Design Description

The Database Design Description (DBDD) describes the design of a data-base, that is, a collection of related data stored in one or more computer-ized files in a manner that can be accessed by users or computer programs via a Database Management System (DBMS). It can also describe the soft-ware units used to access or manipulate the data.

12.9.6 Software Design Description

The Software Design Description (SDD) describes the design of a computer software. It describes the software-wide design decisions, the software architectural design, and the detailed design needed to implement the software. Interface Design Descriptions (IDDs) and Database Design Descriptions (DBDDs) may supplement the SDD, as described below. The SDD, with its associated IDDs and DBDDs, is used as the basis for implementing the software. It provides the acquirer visibility into the design and information needed for software support.

12.10 TEST READINESS REVIEWS

The purpose of the Test Readiness Review (TRR) is to ensure that the software tests are complete and carry out the intent of the software testing plan and descriptions, and that the software to be tested is under formal control and ready for testing. This review will be conducted after software-testing procedures are available and CSC integration testing has been successfully completed.

A TRR meeting between all personnel associated with a test is usually held prior to test initiation. The objective of this meeting is to verify that adequate reviews and preparations have been completed to ensure that all necessary actions have been or will be taken to safely meet the test objectives to obtain a successful test or test series.

The responsible software engineer or delegate will call the TRR and act as chairperson. At a minimum, the following personnel will be requested to attend:

- Responsible team leader or delegate (chairperson)
- Task team members
- Development Engineer for item being tested
- System Safety Engineer (if applicable)
- Software Quality Engineer (if applicable)
- Senior management or delegate

The chairperson will prepare and use as the agenda a briefing chart(s) for each applicable topic. All attendees will sign an attendance sheet. The

Name: _____		Signature: _____		
Position		Date: _____		
Title: _____				

No	TOPIC	Completed		NOTES
		YES	NO	
1.	Has the Approved Software Test Plan been issued?	☐	☐	
2.	Have the Approved Software Test Procedures been issued?	☐	☐	
3.	Have results of unit testing, integration testing, and dry-runs been described?	☐	☐	
4.	Has the list of required test equipment been issued?	☐	☐	
5.	Is the Test environment defined and detailed?	☐	☐	
6.	Has the status of facilities and tools been defined?	☐	☐	
7.	Have the data acquisition, handling and analysis provisions been met?	☐	☐	
8.	Are the required procedure manuals available?	☐	☐	
9.	Have all documents been provided for the review?	☐	☐	

FIGURE 12.1
Test Readiness Review Checklist.

chairperson will issue the minutes of the meeting with assigned actions immediately following the meeting. A copy of the briefing charts and attendance sheet will be filed with the minutes. The test request will be reviewed and all present must agree that the test can be run as requested. All disagreements will be resolved before the test proceeds. Figure 12.1 provides an example of a software TRR checklist. The use of this checklist is optional at the discretion of the software program or project manager.

NOTE: References to analyses and data that support the items in the checklist should be recorded so that the checklist may serve as a design audit trail.

12.10.1 Software Test Plan

The Software Test Plan (STP) describes plans for qualification testing of computer software and software systems. It describes the software test environment to be used for the testing, identifies the tests to be performed, and provides schedules for test activities.

12.10.2 Software Test Report

The Software Test Report (STR) is a record of the qualification testing performed on a computer software system or subsystem, or other software-related item. The STR enables the acquirer to assess the software testing and its results.

12.10.3 Version Description Document

The Version Description Document (VDD) will identify and describe a software version consisting of one or more computer softwares. It is used to release, track, and control software versions. The term "version" may be applied to the initial release of the software, to a subsequent release of that software, or to one of multiple forms of the software released at approximately the same time (for example, to different programs or projects).

12.11 SOFTWARE COMPLIANCE STATUS

Software Compliance Status is documentation that includes a software delivery package indicating the compliance status for all software directives applicable to the software delivery and completion of required software engineering reviews and audits.

12.11.1 Software Development Files

The Software Development Files (SDFs) are hard copy or software and/or electronic files that contain data to be collected during each development project. For each project, the SDFs may consist of multiple component parts, each of which is referred to in an SDF at the appropriate level of structure (usually no more than three levels are required), but contain different levels of information. Each project determines whether to create software-unit-level SDFs or whether to include that information in the intermediate-level SDFs. Software will also have an SDF that summarizes all the information in its lower-level SDFs. Use the simplest SDF structure that meets the needs of the project. Whatever structure allows unit compilation and independent unit-level testing is usually the correct level. The SDF includes the following:

- Software name or title
- Software Unit (SU) or Computer Software Unit (CSU) level SDF(s)
- Intermediate (CSC) level

SDFs are updated throughout the project, as the content items become available from development progress. Most activities include task statements identifying items to include in the SDFs. The SDFs are usually configuration managed under the "informal" level of control. This affords the most flexibility to the development team. The SDF can reference or contain documents, or portions thereof, and data that are under stricter levels of control without compromising the contents because once they are moved into the SDF, they are considered "working papers."

Only if the personnel working at a specific level want to change one of these controlled documents do they have to exercise appropriate control, which often requires a formal change request. Exceptions to this informal level of control can be imposed by raising the SDF to a "Managed and Controlled (M&C) level of understanding." This level enforces additional rules, check-in, check-out, version control, etc., and ensures that larger projects are more rigorously managed to prevent inadvertent or uncoordinated changes in one area that could impact another part of the software.

12.11.2 Software Verification and Validation

This book describes SQA activities that are performed in support of verification and validation. The purpose of verification is to ensure that the product meets its specified requirements, which is accomplished through various activities ranging from in-process Peer Reviews and lower-level testing to qualification activities that may include testing, inspection, analysis, or demonstration. The purpose of validation is to ensure that the product fulfills its intended use when placed in its intended environment. SQA will audit test activities support qualifications by monitoring the activities.

12.12 FIRST ARTICLE INSPECTION

The First Article Inspection (FAI) for software includes verification and assurance that software documentation, software products, and configuration requirements are defined in order to meet military and aerospace requirements, and that the software can be consistently reproduced. The inspection performed will also ensure that software engineering requirements and processes have been applied to development and release activities. The software is reviewed and approved by a Software Design/ Developer Engineer and Quality Assurance.

12.13 PROCESS STEPS

The basic FAI process steps are required on detailed software items. The process steps end when software records are stored using the audit exit criteria. The process flow is described in Figure 12.2.

12.13.1 Determine Levels

The software engineering team will initiate an FAI software decision sheet and coordinate with Quality personnel at appropriate levels. The

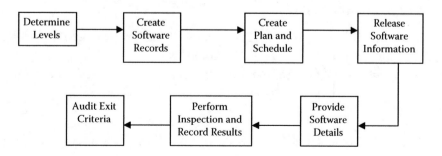

FIGURE 12.2
FAI process flow.

decision sheet will be submitted to Quality personnel for initiation of the FAI.

12.13.2 Create Records

For each CI/CSCI, part number records are created. Document all design characteristics that require FAI verification and validation as required per the required records. Attachments (software engineering drawings or documents, reports, etc.) are to be used and include instructions for their use. Submit the records to engineering for planning and schedules.

12.13.3 Plan and Schedule

Software Engineering and Quality personnel create the plan and schedule to perform the FAI. I work with software suppliers, and our team will request the required documentation at least three weeks before the FAI is kicked off. Some suppliers do not understand the FAI process, so we have to educate them and provide them with a list of what documentation is required for our review. The software supplier will answer and say, "Oh, now we understand."

12.13.4 Release Information

The required released information and plans generated are provided to Senior Management to ensure that the FAI is ready to begin. Senior Management will receive the FAI requirements for review and implementation and its decision to start the FAI will be initiated.

12.13.5 Provide Details

Software Engineering and Quality Assurance will provide detailed processes for performing the FAI. Deviation from the implementation of the processes is not permitted without concurrence between Software Engineering and Quality personnel.

12.13.6 Perform Quality Inspection and Record Results

For software products, Software Engineering and Quality will perform a quality inspection as documented in the FAI software checklist in

Table 12.2. Once all FAI activities are complete, Quality personnel will perform a final FAI acceptance. Software Quality Assurance will review the FAI results documented for acceptance, completion, and apply a signature of approval. When nonconformance issues are found, the software quality will require identification of specific design characteristics, and these will be listed and associated with the FAI.

12.13.7 Software Documentation

The Software Documentation including additional requirements, design, and testing is a significant factor. The traceability of requirements, design, code, and all testing is critical to support the FAI. Every line of object code (not just source code) will be traceable to software requirements.

12.13.8 Audit Exit Criteria

Audit Exit Criteria will list the results of the FAI and collected evidence against the software inspection criteria. Major nonconformities, minor nonconformities, and opportunities for improvement are included.

12.14 FCA/PCA AUDITS

The required audits will be performed on all items with unique development changes in accordance with MIL-STD-1521B Technical Reviews and Audits for Systems, Equipments, the required Configuration Management Plan (CMP), and the Configuration Audit Plan (CAP) (Figure 12.3).

12.15 FUNCTIONAL CONFIGURATION AUDIT

The Functional Configuration Audit (FCA) will verify that the CI/CSCI performance complies with the hardware, software, and the Interface Requirements Specification (IRS). The test data is required to be reviewed and verified knowing that the hardware and software performs as required by the functional and allocated configuration identification. The FCA will be the prerequisite to acceptance of the configuration

TABLE 12.2

FAI Software Checklist

ID	Requirements	Guidance	Y/N	Documentation Reference	Auditors Comments
1	Software Products are controlled in the CPL	Review the SCMP			
2	Software requirements deviations are recorded and approved	Severity codes reviewed			
3	Software Change Requests (SCRs) comply with the SCMP	Inspect the SCM plan			
4	SCRs in this SW release are identified	Review the VDD			
5	Open SCRs not incorporated by this release are identified	Review the VDD			
6	Software releases are traceable to the previous baselines	Record the VDD section			
7	Is the software end item identical to the tested software?	Review test plans			
8	Objective evidence that the software build process was performed to released procedures	Review software build requests			
9	The software media part numbers are identified	Review media and VDD			
10	Objective evidence provided that SW loading of the object code into the target systems was performed by approved and released procedures	Review SW loading plans or procedures			

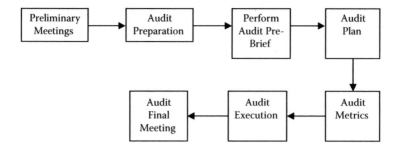

FIGURE 12.3
FCA/PCA process steps.

item. A technical understanding will be accomplished concerning the validation and verification per the Software Test Plan (STP) concerning the software.

FCA activities include the following:

- Verification that the CI/CSCI performs to required configurations
- Release of major or minor engineering changes
- Establishment of a product baseline

12.16 PHYSICAL CONFIGURATION AUDIT

The Physical Configuration Audit (PCA) identifies the product baseline for production and acceptance of the CI/CSCI audited. PCA verifies that the "as-built" configuration correlates with the "as-designed" product configuration, and that the acceptance test requirements are comprehensive and meet the necessary requirements for acceptance of the production unit. Equally important, it demonstrates that management systems for quality, engineering, and configuration management information accurately control the configuration of subsequent production units. Incremental and progressive audits are performed on systems and major assemblies to build up to the PCA. Just as a note: it is an option for the PCA to be conducted concurrently with the FCA, but the norm is to have the PCA conducted and performed after FCA. Extracts from the previous FCA plan will be made available to the team. Quality Assurance (QA) and Senior Management ensure that available budget and engineering personnel are executed per the PCA Audit Plan. Metrics captured for the FCA are similar for the PCA, to be compiled and reviewed during the audit. After the PCA execution and the metrics are completed, the schedule for the PCA final meeting is coordinated with the customer.

PCA activities include the following:

- Verify that the "as-built" configuration correlates with "as-designed" product configuration.
- Determine the acceptance test requirements per Quality Assurance.
- Release Engineering Changes.
- Establish the final product baseline.

12.17 PRELIMINARY MEETINGS

Preliminary Meetings are conducted with Quality Assurance (QA), Configuration Management (CM), and the customer. Separate meetings are conducted with Systems Engineering (SE) to review and ensure that the allocated requirements for the Contract Data Requirement List (CDRL) and Supplier Data Requirements List (SDRL) are completed. The contractual plans and also internal documentation must be updated and released through Data Management (DM). FCA/PCA will include reviewing hardware Configuration Items (CIs). Senior Management will ensure available budget and engineering personnel to execute the FCA/PCA Audit Plan.

12.18 AUDIT PREPARATION

Audit Preparation will ensure that all Analysis Reports and the Acceptance Test Reports (ATPs) are verified and complete. All Specification Compliance drawings and documentation are approved and released. The requirements baseline will be current and all higher- and lower-level changes are incorporated and Deviations and Waivers are approved. The list of precoordination items is as follows:

- Assign QA or CM to chair the FCA/PCA.

- Establish start and end date.
- Assign team members.
- Assure facility for the team is available.
- Prepare agenda.
- Coordinate preliminary agenda with the audit team.
- Provide final agreement of the agenda.

12.18.1 Collect Engineering Data Package

Collection of the Engineering Data Package supports Audit Preparation and provides the necessary data/information to the Audit Team to prepare for the performance of the audit. The Engineering Data Package will comprise the following:

- Requirements baseline (DOORS for System, Segment, and Subsystem) data
- Configuration Status Accounting (CSA) reports
- Internal reconciliation of all Configuration Accountability Verification System errors
- Engineering Software Documents
- Verification Plans, Acceptance Test Procedures, and Software Test Descriptions
- Acceptance Test Reports and Specification Compliance Reports/ Documentation
- Deviations and Waivers
- High- and middle-level changes
- Open paperwork:
  - Nonconformance Reports
  - Test Problem Reports (TPRs)
  - Software Change Requests (SCRs)
- Software products as listed in the Software Development Plan (SDP)
- Drawing/document availability from Supplier/Subcontractor (required for PCA)
- Build records availability from Supplier/Subcontractor (Required for PCA)
- Assemble Engineering Data Package for FCA/PCA.

12.19 PERFORM AUDIT PRE-BRIEF

The audit approach will ensure the development of an audit plan for each item being audited. An audit pre-brief meeting will be held with team members to discuss expectations and responsibilities. If the FCA/PCA includes suppliers or subcontractors, they will work together with the contactor to develop an audit plan to meet requirements applied between both parties. After the FCA/PCA execution and the metrics are completed, the schedule for the FCA/PCA final meeting is coordinated with the customer. Determine the FCA/PCA readiness for a "go" or "no-go" decision.

For software-related items, the SCM organization will ensure that versions of the Interface Requirements Specification (IRS), System Requirements Specification (SRS), test plans/procedures, test reports, Version Description Documents (VDDs) are released and applicable deviations are closed. Computer Software Configuration Items (CSCIs), product specifications, and related documentation are also available for the FCA/PCA audits. The audit team will verify that the software performs its functional/allocated configuration requirements. The Software Test Reports (STRs) will be reviewed for validity and completeness.

12.20 AUDIT PLAN

It is important to have an Audit Plan for organizations to be in sync when the FCA/PCA is being performed. Below is the template to ensure that effective FCAs/PCAs are conducted and performed to defined requirements:

Conduct FCA/PCA
- Conduct in-brief
- Team introduction
- Review audit scope
- Review entry criteria
- Review agenda

Perform FCA/PCA Audit
- Review Engineering Data Packages
- Review minutes daily
- Review action items

TABLE 12.3

FCA/PCA Schedule

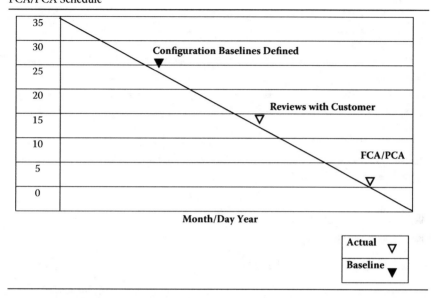

- Prepare FCA certification sheets
- Review final action items
- Prepare minutes
- Review Configuration Audit Summary Report (CASR)

A committed Audit Schedule will be coordinated with the affected organizations and approved by senior management. Refer to Table 12.3, FCA/PCA schedule, for an example that covers 35 days from the start of the FCA/PCA to completion.

12.21 AUDIT METRICS

The Audit Metrics in Table 12.4, FCA/PCA responsibilities, show a defined baseline, scheduled reviews, and the expected start of the FCA/PCA. FCA/PCA metrics will also include responsibilities for establishing the time, place, and agenda and may be subject to coordination with the customer. The FCA/PCA coordination with the customer will be accomplished in advance of each audit to allow adequate preparation for the meeting.

TABLE 12.4

FCA/PCA Responsibilities

Name	System	Task	Status
Engineering, Quality	ID	Design documentation	mm/dd/yy
Configuration Management	ID	Data Package updated	mm/dd/yy
Chairperson	ID	Submit FCA/PCA agenda	mm/dd/yy
Software Organization	ID	Software source code available	mm/dd/yy
Systems Engineering, DM	ID	IRS/SRS complete	mm/dd/yy
Data Management	ID	Applicable documents complete	mm/dd/yy
Chair	ID	Readiness preparation complete	mm/dd/yy
Quality, Engineering, CM	ID	Finalize FCA/PCA data package	mm/dd/yy
Audit Team	ID	Conduct FCA/PCA	mm/dd/yy
Audit Team	ID	FCA/PCA action items	mm/dd/yy
Audit Team	ID	FCA/PCA complete	mm/dd/yy

12.22 AUDIT EXECUTION

The Audit Execution activities require the review to show evidence of requirements analysis, related verification planning, released plans, and drawings or documents showing results. There are discussions and reviews of data relative to specification qualifications, documents, data, reports, etc., required by contract. Test and analysis data will be reviewed to ensure that allocated requirements are completed and released. Also, support documentation is reviewed. Preparation of a summary report identifying action items and responsibilities for resolution will be initiated during the audit. The summary report will be provided to all affected program or project support organizations.

12.23 AUDIT FINAL MEETING

The Audit Final Meeting involves the review and concurrence of the baseline. Configuration audits for hardware/software include the witnessing/monitoring by the Program and Prime QA personnel of development/testing of designated CI and CSCI end items. The buy-off of inspection, audits, and maintenance include the following:

- Waivers, deviations, and change status

- Reviews of the development of CI and CSCI specification compliance
- Reviews of drawings and software part documents for product specifications
- Certification of engineering releases and quality evaluations
- Capability of controlling configuration of subsequent units

The FCA/PCA for a CI and CSCI final meeting may be co-chaired by the customer. At the meeting, satisfactory completion status of the FCA/PCA tasks is verified, the acceptance test procedure is approved, and the change incorporation and waivers or deviation status is reviewed. The verification analysis/inspection reports problems, if any, which are documented as action items, and the FCA/PCA Certification Sheets are signed. A closure letter is prepared and submitted to Contracts after the last action item is closed out to close the FCA/PCA.

12.24 AUDIT METHOD FOR SOFTWARE

A Functional Configuration Audit (FCA) and Physical Configuration Audit (PCA) method for software is performed to confirm that the software item under audit conforms to the relevant engineering documentation, processes, and quality assurance requirements that have been tested in accordance with the approved Acceptance Test Procedures (ATPs). As part of the FCA audit process, the engineering release and change procedures are examined to ensure that they adequately control the authorized configuration and that approved changes are effectively implemented as intended. The FCA is used to establish the Product Configuration Baseline. The FCA/PCA for software is accomplished using an incremental audit approach. MIL-STD-1521b provides general guidance for conducting and performing audits.

Comment: Here is where there is disagreement within programs or projects on who should conduct an FCA/PCA for software. I feel that Software Quality Assurance (SQA) should conduct and perform the FCA/PCA for software. Again, SQA, step up to the plate and don't watch from the bench. These two audits for software will need to be defined in the Software Development Plan (SDP), Software Quality Assurance Plan (SQAP), Software Configuration Management Plan (SCMP), and the Configuration Audit Plan (CAP). Closing comments: Participating in FAI, FCA, and PCA preparation meetings has been an adventure. Questions are asked to IPT team members and Senior Management about starting contractual formal audits. You know what the answers are? No! When is the formal audit scheduled to start and are we ready? No! What? Did we perform Internal/ Informal Software Engineering Reviews and Audits? No! If you schedule and perform these Internal/Informal Engineering Reviews and Audits for software, you will hear, "Are we prepared and ready to perform these formal audits?" Yes! Let's get going. I guarantee success and the customer(s) you serve will be thrilled at the end!

12.25 FINAL SUMMARY

Assigned personnel are required to understand project management concepts and the performance of software engineering review and audit activities. The customer will request daily updates regarding the quality of project plans, procedures, implemented software development processes, and software work products required to support critical company deadlines and program schedules. These reviews and audits for software provide validation and verification of delivered software products and the assurance required for proper installations in software development labs, system integration labs, and customer deliveries. Performing the incremental engineering reviews and audits will minimize problems that are witnessed and discovered during contractual allocated inspections and formal configuration audits.

FURTHER READING

Johnson, Spencer. *Who Moved My Cheese*, New York, NY: Penguin Putnam, 1998.
Keyes, J. *Software Configuration Management*, Boca Raton, FL: Auerbach Publications, 2004.
MIL-STD-480 Configuration Control: Engineering Changes, Deviations, and Waivers.
MIL-STD-973, Configuration Management.
MIL-STD-1521 Technical Reviews and Audits for System, Equipment, and Computer Software. "This Military Standard is approved for use by all Departments and Agencies of the Department of Defense" (DOD).

Appendix A:
Acronyms and Glossary

The terms associated with the Software Engineering Review and Audit process described in this book are defined. Where a definition source is supplied, it is placed in "( )"; the order of reference is (1) MIL-STD (2) CMMI (3) QMS, (4) SQA (5) SEMP (6) SDP, (7) (PAT), (8) SCM, (9) SOW, (10) IPT, (11) CA, (12) FAI, FCA, PCA, or other common usage in the book.

Acceptance criteria: The criteria that a system or component must satisfy in order to be accepted by a user, customer, or other authorized entity.

Audit: An independent examination of a work product or set of work products to assess software compliance with specifications, standards, contractual agreements, or other criteria.

Baseline: A specification or product that has been formally reviewed and agreed upon and can only be changed through formal control processes.

Build: Operational version of a software product incorporating a specified subset of capabilities that informal and formal products will include.

Certification: A written guarantee that a system or computer program complies with its specified requirements.

Change Control: The processes by which a change is proposed, evaluated, approved or rejected, scheduled, and tracked.

CMMI: Collection of process models and methods for use in new disciplines to be integrated for organizational structures.

Code and Unit Testing: A routine is written and the disciplines in which data will be represented are specified.

Computer Data: Data available for communication between or within computer equipment.

Computer Language: A defined structure devised to simplify communications with a computer.

Computer Program: A sequence of instructions suitable for design, coding, and authorized processing per a computer system.

Computer Program Library: A Software Library providing permanent archival storage for software and related documentation.

Computer Software Component: A logical or functional grouping of CSUs to which the SCM tools assign a unique name, which is supplied by the software designer.

Computer Software Configuration Item: An aggregation of software designated for configuration management and control.

Computer Software Units: File names that consist of a three-character CSCI abbreviation, followed by a three-character CSC abbreviation and a descriptive CSU name (abbreviated to meet operating system and character limits) plus a version number.

Configuration Audit Plan: A plan that is used as configuration audit steps for internal and formal audits being performed by a contractor.

Configuration Item: Configuration items in systems are designated and their characteristics are recorded.

Configuration Management: The process of identifying and defining the configuration items in a system, controlling the changes and release of these items throughout the system lifecycle, and recording and reporting the status of change requests to verify completeness.

Configuration Status Accounting: The recording and reporting of information that is needed to manage a configuration including a list of approved changes and documentation.

Control Files: Files stored in a computer that are controlled and password protected.

Critical Design Review: A formal meeting at which the critical design is presented to the user or customer for comment and approval.

Data: A representation of facts, concepts, or instructions suitable for communication, interpretation, or processing.

Data Management: A function of controlling the acquisition, analysis, storage, retrieval, and distribution of data.

Defect: A condition that causes software development, coding to fail, and perform its required function.

Delivery: The point in the software development lifecycle at which a product is released to its user for operational use.

Design: The purpose of defining the software architecture, components, modules, interfaces, and data for a software system to satisfy specified requirements.

Design Phase: The period of time in the software lifecycle for software development.

Documentation: The collection and management of documents identifying plans, processes, and procedures.

Drawing: A computer depiction of graphics or a manually prepared graphic representation of a part or product.

Engineering Review Board: This is established for the software Integrated Product Teams (IPTs) to review and disposition changes that affect controlled software and related documentation.

File Name: A name given by a software designer to a specific collection of data.

Firmware: A computer program stored in a hardware unit as an integrated circuit with a fixed configuration that will satisfy a specific software application or operational requirement.

First Article Inspection: The inspection performed to ensure that software engineering requirements and processes have been applied to development and release activities.

Formal Testing: The process of conducting testing activities and reporting the results in accordance with approved test plans.

Functional Configuration Audit: Prerequisite to acceptance of the configuration item. A technical understanding will be achieved in regard to the validation and verification per the Software Test Plan (STP) concerning software.

Hardware: Physical equipment used in data processing, as compared to computer programs, plans, procedures, and associated documentation.

IEEE: Institute of Electrical and Electronic Engineers, which is accredited to ANSI standards.

Implementation Phase: The period of time in the software lifecycle during which software products are created from design documentation.

Inspection: A formal evaluation in which software requirements, design, or code is examined in detail to detect faults, violations of development standards, and other problems.

Integration Testing: An orderly progression of testing in which elements of software and hardware are combined and tested.

Interface Requirement: A requirement that specifies a hardware, software, and database with which a system must interface.

IPT: The Integrated Product Team is accountable and is composed of members from the appropriate functional disciplines (e.g.,

Engineering, Supplier Management, Product Support, etc.) necessary to accomplish day-to-day activities.

Item: An element of a set of data, such as digits, bits, or characteristics, treated as a unit.

Modification: A change made to software and the process for that change.

Nondevelopment Item: Software used to assist in the development of the deliverable CSCIs, but not identified as a deliverable product.

Object Code: The output from a compiler that is directly executable by the computer system.

Physical Configuration Audit: Identifies the product baseline for the production and acceptance of the CI/CSCI audited. PCA verifies that the "as-built" configuration correlates with the "as-designed" product configuration and that the acceptance test requirements are comprehensive and meet the necessary requirements for acceptance of the production unit.

Preliminary Design Review: A formal meeting at which the preliminary design is presented to the user or customer for comment and approval.

Procedure: The documented description of a course of action taken to perform activities or resolve problems. Manual steps or processes to be followed.

Process: To perform to defined instructions during the software development lifecycle.

Program: A schedule or plan that specifies actions to be taken.

Project Plan: A management approach that describes the work to be done, resources required, methods to be used, reviews, audits, the configuration management, and quality assurance procedures to be implemented.

Qualification Testing: Formal testing conducted by the developer for the customer to demonstrate that the software meets specified requirements.

Quality: The totality of features and characteristics of a product or service that has the ability to satisfy required needs.

Quality Assurance: A planned and systematic pattern of all actions necessary to provide confidence that the product conforms to established technical needs.

Quality Management System: Software industries and software programs that establish, document, implement, and maintain effective quality management and will continually improve their effectiveness.

Quality Metrics: Measurement of the degrees to which software possesses given attributes that affect quality.

Requirement: A condition or capability needed by a user to solve a problem or achieve an objective. The condition or capability must be met by a system to satisfy a contract, standard, or specification.

Requirement Analysis: The process of studying user needs to arrive at a definition of system or software requirements. Verification is also performed for systems and software requirements.

Requirements Phase: The period of time in the software lifecycle during which the requirements of a software product, such as functional and performance capabilities, are defined.

Review: Informal or formal review of system requirements, software design, software configuration management, software quality, testing, and required data to show compliance with documented plans, processes, and procedures.

Risk Management: A process to identify risks and approaches to prevent future risks.

Software: Computer programs, procedures, rules, and any documentation pertaining to the operation of data processing systems. It is in contrast with Hardware.

Software Build Request: Requests to Software Configuration Management to provide software builds for software systems and computer labs to support a Formal Test.

Software Configuration Management Plan: Configuration Management Plan for the control and management of software products during a software development program.

Software Configuration Management Plan: Configuration Management Plan for the control and management of software products during the phase of a software development program.

Software Development Plan: Establishes the plan for the development of software during the lifecycle of a program.

Software Development Process: The process by which the requirements allocation process addresses software activities to receive, analyze, and allocate system level design requirements to be certified for operational use.

Software Documentation: Technical data or information that describes or specifies the design or details, explains the capabilities, and provides instructions for using software.

Software Engineering: A systematic approach to the development, operation, and maintenance of software development.

Software Engineering Environment (SEE): The design, installation, operation, maintenance, and configuration management of software engineering assets.

Software Engineering Institute: Resources for improving management practices for addressing software and disciplines that affect software.

Software Lifecycle: The period of time that begins with the decision to develop a software product and ends when the product is delivered.

Software Load Request: A request provided to Software Configuration Management to build and load software in systems for quality verification and verification by Quality Assurance.

Software Maintenance: Modification of a software product after delivery.

Software Product: A software entity designated for delivery to a user.

Software Quality: Features and characteristics of a software product that satisfy a need and conform to specifications.

Software Quality Assurance: A planned and systematic approach to provide adequate confidence that the product conforms to established requirements.

Software Tools: Computer tools used to develop, test, analyze, and maintain a computer program and its documentation.

Source Code: Computer programs written in a computer language that requires a translation provided by a computer system.

Statement of Work: Processes and procedures supporting the work defined by a purchase contract including the technical areas.

Subsystem: A group of assemblies or components or both combined to perform a single function.

Supplier Data Requirements List: Track specification control documents, supplier's design, approvals, and acceptance.

Systems Engineering: Analysis, requirements understanding, and the importance of software design capabilities. Interfaces are defined externally and internally to ensure hardware and software is compatible and supporting team activities.

Test Readiness Review: Ensures that the software tests are complete and carry out the intent of the software testing plan and that descriptions and software to be tested are under formal control and ready for tesing.

Test Report: A document describing the conduct and results of testing carried out for a system or system component.

Testing: The process of exercising or evaluating a system by manual or automated means to verify that requirements satisfy expected results.

Validation: Validation demonstrates that the product, as provided, will fulfill its intended use.

VDD: Identifies and describes a software version consisting of one or more computer software. It is used to release, track, and control software versions.

Verification: Verification addresses whether the work product properly reflects the specified requirements.

Waiver: A written authorization to accept a software configuration item or other designated item which, during production or having been submitted for inspection, is found to depart from specified requirements, but is nevertheless considered suitable for use as is or after rework by an approved software method.

Appendix B:
Software Development Plan

**DOCUMENT
NUMBER:**

REVISION:

REVISION DATE:

SDP Information

Release Date	

Signatures for SDP Approval and Release

AUTHOR: _____   _____   _____
                          Signature   Org.   Date

APPROVAL: _____   _____   _____
                          Signature   Org.   Date

DOCUMENT
RELEASE: _____   _____   _____
                          Signature   Org.   Date

Contents

List of Figures

List of Tables

Abstract

This document describes the software plan and approach for developing the software.

B.1 OVERVIEW

The Software Development Plan (SDP) describes and identifies the software development methodology, the organizations responsible for performing the software development and maintenance, the procedures and activities these organizations follow, and the resources to be used for software activities. The SDP is the governing software planning document.

B.1.1 Purpose, Scope, and Objectives

This section contains a full identification of the software to which this document applies, including software end item identification information. Any simulations or support software items to be developed to aid in software- or system-level testing and analysis should be listed in this section. A table may be used to list the software end item information.

This Software Development Plan (SDP) details the software development plans and procedures for both mission-critical and non-mission-critical software. Supplier software development plans are derived from this plan and comply with it. This SDP includes the management guidelines applied to all software developed, tested, and maintained during the lifecycle, including supplier and team deliverables. The software development methodology is used and applied for the production of the software. The institutionalized and tailored process is the basis for the externally assessed SEI-Level rating. Supplier Quality surveys suppliers as needed to approve their quality system and to maintain approval ratings in the Approved Supplier List. Software Quality Engineering monitors supplier performance, and verifies that the software quality system implemented is effective and compliant to applicable requirements. The supplier provides evidence of CMMI appraisals and ratings as required by contract.

This SDP focuses on the set of software engineering processes and the work products (e.g., specifications, documents, test plans, and executables) that flow to and from the processes, to management, and eventually to the customer. It includes the schedule and organization to implement software engineering processes. The SDP satisfies the requirements for the Contract Data Requirements List (CDRL) and Statement of Work (SOW). The detailed purpose and function of the system is described in the operational concept document, or System Engineering Management Plan (SEMP). The system consists of all of the equipment, facilities, software, and support functions necessary to achieve the specified mission. Other plans describe the overall organization structure supporting the development of the components. Organizational areas covered by plans to include management, system engineering, hardware and software engineering, and engineering support, testing, and evaluation, as well as deployment and sustaining engineering support.

B.1.2 Assumptions, Requirements, and Constraints

This section describes the assumptions on which the project is based. It also includes any constraints on the project that affect the schedule, budget, resources, process maturity level, software to be reused, customer software to be incorporated, technology to be employed, and interfaces to other products.

The system requirements are documented and flowed down from multiple sources including the System/Segment Design Document, the Component Specifications, and the System Specification.

B.1.3 Schedule, Budget, and Resource Constraints

Schedules are captured and described in Section B.2.3, Software Schedules. Build and release plans are developed to meet specific milestones.

B.1.4 Software Process Maturity Constraints

Software development will apply software engineering processes and methods to be consistent with the Capability Maturity Model Integration (CMMI)-SE/SW/IPPD/SS V1.2 (CMMI for Systems Engineering, Software Engineering, Integrated Product and Process Development, and Supplier Sourcing) to achieve Level 3 or higher.

B.1.5 Evaluating and Incorporating Reusable Software Products

This section describes the approach to be followed for identifying, evaluating, and incorporating reusable software products, including the scope of the search for such products and the criteria to be used for their evaluation. Reusable software products may include software from existing products, Commercial Off the Shelf (COTS). Candidate or selected reusable software products known at the time of preparation or updating of this plan are identified and described, together with benefits, drawbacks, and restrictions, as applicable, associated with their use. Each source of reusable software is evaluated on the characteristics listed below, as applicable, to determine whether the reusable software satisfies the needs:

- Ability to provide required capabilities and meet required constraints
- Ability to provide required safety, security, and privacy
- Reliability/maturity as evidenced by established and documented records
- Testability
- Interoperability with other system and system-external elements
- Fielding issues, including restrictions on copying and distributing the software, or documentation and license or other fees applicable to each copy
- Maintainability, including the likelihood that the software product needs to be changed, the feasibility of accomplishing that change, the availability and quality of documentation and source files, and the government data rights to the software product
- Short- and long-term cost impacts of using the software product
- Technical, cost, and schedule risks and trade-offs in using the software product
- Availability of artifacts to support certification requirements

The identification of COTS is provided in this section as well as the approach for acquiring these products and the criteria for accepting them as part of the software end item. Each software item selected for reuse is evaluated to determine if it meets the system requirements and the expected fidelity. Software items that are fully compliant with the requirements are reused without modification. Software items requiring modification are evaluated to determine the amount of change required. All software is integrated and tested during formal software testing. Lower-level testing is performed to the extent that the reusable software has been modified.

B.1.6 Developing Reusable Software Products

All developed software is considered for inclusion in a reusable library for future software development projects, where release restrictions and license arrangements permit. Future updates to software will comply with the desire for flexibility and growth potential in support of system life-cycle maintenance. The primary supportability features in the design are as follows:

- Modularity of design
- Provisions for processor memory and duty cycle reserves
- Use of layered software architectures that isolate the application software from the processor hardware dependencies

Software is developed using the design methods specified in Section B.3.4, Software Design, to decompose the software into modular elements. The software coding standards preclude platform-dependent software design to the maximum extent practical and to keep the amount of platform-dependent software to a minimum. The modular design and the platform independence provide a structure in which the developed software can be reused Software Products. This section defines the software work products (i.e., documents) to be produced as required by contract (i.e., Contract Deliverable Items List [CDRL]). It also includes any nondeliverable software work products. Software work products listed in Table B.1 that are not required by the contract or that will not be produced are removed from the table. Additional software work products may be added to the table as required by the contract. External documentation is prepared, submitted, and approved in compliance with the Contractor Deliverable Requirements List (CDRL) and its specific data item description. Internal documentation is prepared and internally approved in compliance with the standards of the contractor format.

B.1.7 Evolution of the Software Development Plan

This section describes the conditions under which the SDP is updated and how changes to the SDP are approved and implemented, including software quality involvement. It describes the plans for adherence to the SDP and the approach for updating the SDP after customer approval. Any updates to plans are subject to customer approval as defined by the contract.

The project management and staff use the SDP to guide the software development activities. The SDP is updated as the contract or software commitments are refined or changed. Each revision of the SDP is reviewed and approved by the project software engineering, software quality engineering, management, and functional management.

B.1.8 Relationship to Other Plans

This section describes the relationship, if any, of the SDP to other project management plans.

This SDP contains the plans pertaining to software development for the software, as flowed from the level plans. The contents of the SDP are flowed vertically, and the Systems Engineering Management Plan (SEMP) describes and depicts from them the hierarchy of the plans. The SDP is aligned horizontally to the other plans such as

- The Risk Management Plan (RMP), which provides for the identification, analysis, assessment, and mitigation of risk issues that could potentially impact the cost, schedule, and performance goals. Additionally, the plan discusses the relationship between Technical Performance Measures (TPMs) and the Risk Management Process.
- The Software Configuration Management Plan (SCMP), which describes the software configuration management plans and activities for the component software end items. It details the processes to be followed for identification, change tracking, release, and delivery of the software end items.
- The Software Quality Assurance Plan (SQAP), which describes the software quality assurance plan and activities for the software end items. It includes the plans for performing and recording the results of software quality evaluations of the software development activities, processes, and resulting software work products.
- The Software Test Plan (STP), which provides the plan for formal qualification testing of the software end items. It includes the software configuration(s) to be tested, the test environment, the regression testing approach, and the plan for tracking anomalies to resolution.

- The System Engineering Management Plan (SEMP), which describes the system engineering management plan and activities for the system. It describes the processes to be followed for the analysis, including trade studies, to derive the system requirements and to perform verification of the system requirements.

B.2 PROJECT ORGANIZATION

This section is divided into subsections that describe the project organization and resources to be applied to the software development activities. Organization Structure is defined in Figure B.1.

To successfully develop the software products, the software team must coordinate with other teams. The software team is a cohesive team of customer and supplier representation to streamline the development of the software products and support the other teams. Management is informed of the weekly software development status, monthly status, and performance and metrics at the monthly reviews. The software team supports management action items, customer reviews, and all major reviews. The software team has the responsibility, authority, and accountability for all software development as described to include the following:

- Planning the overall software development
- Executing and monitoring software development efforts through requirements, design, code, test, and delivery
- Coordinating the development and delivery of all software products across the individual software development teams

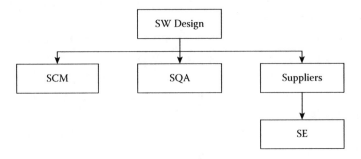

FIGURE B.1
Organization structure.

- Coordinating with other disciplines (e.g., systems engineering, hardware/software integration testing and evaluation, and hardware development) to ensure that software is considered as design decisions and trade-offs are made
- Coordinating interfaces between software elements of the system and between hardware and software
- Collecting and managing to software metrics
- Gathering status and plans, agreeing on commitments, surfacing issues and risks, resolving conflicts, and facilitating a team environment
- Involving the customer as an active participant in developing software
- Coordinating on-site support at supplier facilities as needed
- Participating in working groups as required

B.2.1 Software Team Roles and Responsibilities

The software team roles and responsibilities, as defined include the following:

- Support to systems engineering in the development of system and subsystem specifications, the system and subsystem design document, and interface control documents.
- Support for the security architecture development, lifecycle cost and cost as an independent variable (CAIV analyses), and system engineering–led trade studies and analyses.
- Coordination with the hardware team in the identification and documentation of hardware/software interfaces.
- Providing integrated and tested software builds to the integration testing and evaluation team to support-level testing.
- Interfacing with the product support team for human factors requirements and assessments, safety requirements and analyses, and maintainability requirements and analyses.
- Interfacing with operations for all level CM, including all formal document and code releases. SCM closely coordinates software deliveries with the CM team.
- Supporting the risk and opportunity management team, participating in the working group and management boards, and actively identifying, documenting, and mitigating software-related risks and opportunities.

- Interfacing with the support team by supporting software quality product evaluations and process audits. Software quality oversight is provided through the quality engineering team.
- Supporting the change management team with software cost, schedule, and technical impacts to proposed changes.
- Working with the site support team to resolve personnel issues, facilities, and resources.

B.2.2 Software Management Process Plans

This section is divided into subsections that describe the project software management plans for the software development project: software schedules, effort, estimation, training, metrics, risk management, and decision analysis/resolution.

B.2.3 Software Schedules

This section describes the software development schedule and provides or references an example schedule. The example schedule identifies the software development activities for each software build and includes the initiation and completion of each activity, dependencies between activities and upon external inputs, and milestones representing software work product completion.

The detailed software schedules are also incorporated into an integrated software schedule, which is discussed and updated during software team meetings and management reviews.

B.2.4 Software Effort

This section describes the personnel resources to be applied to the project. It will include, as applicable, the following:

- The estimated staffing plan for the project (number of personnel over time)
- The breakdown of the staffing plan numbers by responsibility (for example, management, software requirements, software design, software implementation, software testing, software configuration management, software quality engineering)

- A breakdown of the skill levels, geographic locations, and security clearances of personnel performing each responsibility

The project software manager is responsible for preparing and maintaining the estimated staffing plan for the project (number of personnel over time), the breakdown of staffing by responsibility and the breakdown of skills required. In addition to the project software manager, software leads are designated to the project. Software projects are staffed with a list of software roles required for software development activites. Software test engineers are identified to ensure that an independent test group performs the testing of the software being qualified. Software project staffing requirements and current levels are reported at scheduled reviews. The assessment of qualifications is considered upon both an academic and experience basis. It is critical that the software staff have experience in numerous software development tools and computer languages. In addition, the project software manager will identify the facilities, equipment and other resources required by the project.

B.2.5 Software Estimation

The software estimation plan describes the method for estimating the software schedule, effort, and other resources necessary for the software development activities. Estimation of any reused software, including the use of COTS and GFE, is also described and addresses follow-up maintenance and support of the products.

The software cost estimate, including the cost for software development and test environment, is updated as the contract or software commitments are refined or changed. Each revision of the software cost estimate is reviewed and approved by the project's software engineering, software quality engineering, management, and functional management.

B.2.6 Software Training

The training plan for software personnel describes the training necessary for the software development personnel to perform the development activities.

As part of the commitment to maintaining the SEI CMMI Level 3 common process maturity level, common process and software engineering trainings are established and available. This training includes courses related to all areas of software development and is required for all personnel involved in the software development process.

B.2.7 Software Metrics Collection

This section identifies the software metrics to be used by software management to aid in managing the software development process and communicating its status to the customer. The software metrics are identified and defined in this section, including the data to be collected, the methods to be used to interpret and apply the data, and the planned reporting mechanism.

The software development organization manages the software development activities using the metrics. The analysis, collection frequency, and thresholds are identified in the Measurement and Analysis Plan. Using software development metrics, software management and software engineers monitor the software development progress and ensure compliance. The Software Metrics Report is developed and delivered.

B.2.8 Software Risk Management

This section describes the approach for performing risk management throughout the software development process. It includes the identification, analysis, and prioritization of the areas of the software development project that involve potential technical, cost, or schedule risks. It includes the plan for developing strategies for managing software risks, recording the risks and strategies, and implementing the strategies in accordance with the plan.

Software risk management activities are performed in accordance with the Risk Management Plan (RMP). The RMP provides for the identification, analysis, assessment, and mitigation of risk issues that could potentially impact the software cost, schedule, and technical performance goals. During each phase of the software development lifecycle, a risk assessment of the software is performed. Approved risks are assigned to the software organization, and the software point of contact provides the risk mitigation status and reporting to software management.

B.2.9 Software Decision Analysis Resolution

This section describes the approach to be followed for recording key decisions made on the project including the rationale for the decisions. The rationale to be recorded may include trade-offs considered, which are documented using trade studies, analysis methods, and the criteria used to make the decisions. The rationale may be recorded in documents, software development folders, code comments, or other media.

Software decision analysis activities are performed in accordance with the System Engineering Management Plan (SEMP). The SEMP describes the process for the analysis, assessment, and capture of all key decisions, including those related to the software development activities. The software organization identifies the areas in which key decisions must be determined.

B.2.10 Software Development Lifecycle

This section describes the software development lifecycle as well as the planned software builds, their objectives, and the software development activities to be performed in each build. If different builds or different software on the project require different planning, these differences are noted as well as any applicable risks/uncertainties and plans for dealing with them. A generic software development lifecycle may be selected, replaced with a representation and implemented for use.

The software lifecycle is also integrated with the lifecycle and the software processes. The number of software builds and the capabilities and schedules of each release support the schedule and development plan as documented in other plans. Each major build may be followed by additional planned or customer-directed releases, which contain software problem fixes, Commercial-Off-the-Shelf (COTS) software upgrades, and minor capability enhancements that do not require significant additional training.

B.2.11 Software Development Environment

This section describes the requirements for the software engineering development environment in order to perform the software development activities. The plan ensures that each element of the environment performs its intended functions in support of the software development activities. This section also describes the plan requirements for the software test environment to perform software testing, including software qualification testing. The plan ensures that each element of the test environment performs its intended functions in support of the software test activities.

None of the software used for designing, building, or testing the software products is deliverable. Any nondeliverable software upon which the operation and support of the deliverable software depends, after delivery, will be identified and provisions made to ensure that the project sponsor/ stakeholder has or can obtain the same software and software products. This includes software used for testing, such as test harness, scripts, test input files, and drivers. Hardware and software items are installed and placed under configuration management. When upgrades or new versions become available, the organization evaluates and recommends whether the update should be incorporated. Upgrades are installed as soon as is reasonable for the development activities if the update is agreed to by all affected organizations. The criteria for evaluating an update include considering problems detected with an update, problems solved by an update, and the impact on the software development effort.

B.2.12 Software Development Facilities

This section includes or references an overview of developer facilities to be used, including geographic locations in which the work will be performed, facilities to be used, and secure areas and other features of the facilities as applicable to the contracted effort. Customer-furnished equipment, software, services, documentation, data, and facilities required for the contracted effort, and a schedule detailing of when these items will be needed will also be included. Other required resources, including a plan for obtaining the resources, dates needed, and availability of each resource item are included as well.

The engineering development teams are primarily located in the internal software development geographic location.

B.3 SOFTWARE LIFECYCLE PHASES

This section contains the detailed software development process(es) to be used in each software lifecycle phase. If different builds or different software on the project require different planning, these differences are noted. The discussion of each activity will be applied to the following:

- The analysis or other technical tasks involved

- The recording of results
- The preparation of associated deliverables, as applicable

Any manual or automated tools and procedures to be used in support of the software methods are also described. The discussion also identifies applicable risks/uncertainties and plans for dealing with them. The following subsections also address the plans for integrating supplier software into the software end items. Any activities in the following sections that are not required may be replaced or removed.

B.3.1 System Requirements and Concepts

This section describes the approach to be followed for software participation in defining the system requirements and concepts. It describes how the software organization participates with systems engineering in analyzing user inputs provided by the customer to gain an understanding of user needs. This section also describes how the software organization participates in defining and recording the requirements to be met by the system and the methods to be used to ensure that each requirement has been met. The resulting information includes all applicable items in the system requirements document, which is produced by systems engineering.

The system requirements definition involves identifying and refining the system-level requirements. The software organization supports systems engineering in defining the system requirements. The system requirements are recorded and tracked. The SEMP or other plans provide the detailed plan for defining the system requirements and concepts. As the software development proceeds through system design and software requirements analysis, the software requirements traceability is expanded and updated. If changes are necessary for any requirement, all higher and lower requirements impacted are identified and analyzed to determine the need for modification.

B.3.2 Software Architecture

This section describes how the software organization defines and records the software architecture definition associated with the software system. The definition includes the software's behavioral design and other decisions affecting the selection and design of software components. The plans for identifying key architecture requirements, constraints, and quality attributes are described. Information pertinent to software architecture and traceability may be captured in a tool(s) used to develop the software architecture and define the traceability.

The software architecture development process involves conducting trade studies and prototypes in order to determine the software architecture. Initially, various software architectures are proposed and examined. The software architecture is developed in parallel with the system architecture development and software requirements definition activities. Traceability of the software architecture components to the software requirements is captured. Reviews of the software architecture work products are described in Section B.6.3, Software Review Plan, and are performed in order to determine the quality and maturity of the software architecture before the software design activities begin.

B.3.3 Software Requirements Analysis

This section describes the approach to be followed for software requirements analysis. Software requirements development is the initial phase in the software development lifecycle. Software requirements analysis is performed to determine new requirements for new software and requirements changes for modified software. Software requirements, including interface requirements, are developed in accordance with the process defined in 07-10003, Software Requirements Development. Software requirements are derived from the system requirements using the methodology, tool, and technique used to develop the software requirements including a brief description of the software's intended use.

B.3.4 Software Design

This section is divided into subsections that describe the approach to be followed for software design. The design pertaining to databases is included as well.

B.3.5 Software Preliminary Design

This section provides the approach to define and record the preliminary design of each software end item (identifying the software units comprising the software end item, their interfaces, and a concept of execution among them). It also describes the approach for traceability between the software units and the software requirements.

Software design activities begin after the software requirements have been identified. Software preliminary design is performed in order to determine the software components and preliminary interface design. The software preliminary design is developed using the software architecture including a brief description of its intended use.

B.3.6 Software Detailed Design

This section describes the approach for developing and recording a description of each software unit comprising the software end item.

Software detailed design activities begin after the software architecture has been completed or when incremental portions of the software end item have been designed (i.e., a function, group of tests, etc.). Software detailed design is performed in order to determine the software and software interface detailed design changes for modified software as well as new software. The software detailed design is developed in accordance with the Software Design Method. The software detailed design is developed using tools and techniques for developing the software design including a brief description of its intended use. Traceability of the software components to the software requirements is captured using tools or techniques used to establish traceability including a brief description of its intended use. Reviews of the software design work products are performed in order to determine the quality and maturity of the software design before the software coding begins.

B.3.7 Software Coding

This section describes the approach to be followed for software coding. If an automated software code generation system or other tools to aid in the software coding are specified then the use of the tool(s) is described in this section. Coding standards to be applied during the development of the source code are included or referenced and include the following:

- Standards for format (such as indentation, spacing, capitalization, and order of information)
- Standards for header comments (requiring, for example, name/ identifier of the code; version identification; modification history;

purpose; requirements and design decisions implemented; notes on the processing (such as algorithms used, assumptions, constraints, limitations, and side effects); and notes on the data (inputs, outputs, variables, data structures, etc.)

- Standards for other comments (such as required number and content expectations)
- Naming conventions for variables, parameters, packages, procedures, files, etc.
- Restrictions, if any, on the complexity of code aggregates

Software coding begins after the software design of a component has been completed. The software coding is developed using tools or techniques for developing the software code including a brief description of its intended use. The coding standards, including the format of the source code, will be applied during the development of the source code per the software coding standards. Traceability of the software code to the software requirements is captured using tools or techniques used to establish traceability including a brief description of its intended use. Reviews of the software code work products are performed to determine the quality and maturity of the code before the software testing begins.

B.3.8 Software Unit Test

This section describes the approach to be followed for software unit testing. If any tools are used to aid in the software unit testing then the use of the tool(s) is described in this section.

Software unit testing begins after the code has been completed. The software unit tests are prepared and conducted using tools or techniques for developing and executing the software unit tests including a brief description of its intended use. Reviews of the software unit are performed in order to determine the quality and maturity of the code before the software integration testing begins. Revision of the software and subsequent retesting is planned and performed based on the analysis of anomalies uncovered during the test.

B.3.9 Software Integration and Integration Testing

This section describes the approach to be followed for software integration and integration testing. If any tools are used to aid in the software integration testing then the use of the tool(s) is described in this section. The approach also addresses the integration and testing of COTS.

Software integration and integration testing begins after the unit testing has been completed. The software integration and integration testing is defined in accordance with the process Software Integration and Integration Testing and as specified in the Software Test Plan. The software integration tests are prepared and conducted using tools or techniques for integrating and testing the software. Software integration testing documentation is provided and performed in order to determine the quality and maturity of the code before the software formal testing begins. Refer to the Software Test Plan (STP) for further information related to the software integration and testing activities.

B.3.10 Software Formal Test

This section describes the approach to be followed for software formal testing. This section defines what tools are used to aid in the formal software testing as well as the use of the tool(s).

Software formal testing begins after integration and integration testing have been completed. The software formal testing is defined in accordance with the process defined in the Software Test Plan (STP). The software formal tests are prepared and conducted using tools or techniques for developing and executing the software formal tests including a brief description of its intended use. Traceability of the software tests to the software requirements is captured using tools or techniques for establishing traceability including a brief description of its intended use. Reviews of the software formal test work products are performed in order to determine the quality and maturity of the code before release of the software for system testing or delivery to the project sponsor and stakeholders. Software test engineers who are independent from the software development engineers perform the software formal testing. Problems detected during the software formal testing are captured and resolved as described in the Software Configuration Management Plan. Refer to the Software Test Plan for further information related to the software formal testing activities.

B.3.11 Hardware/Software Integration and System Testing

This section describes the approach to be followed for software support to the hardware and software integration and system testing. It describes how the software organization participates in developing, recording, and executing the integration system tests as well as the test result analysis and any necessary retesting. The tests cover all aspects of the system-wide and system architectural design. Software and hardware end item integration and system testing involves the integration of multiple software and hardware end items and testing of the integrated system. The end items are integrated and tested in an orderly, planned, and iterative process. This plan is developed as part of the preparation activities and describes the integration and test environment, the tasks, and the schedules for performing the integration and tests. Dependencies between the software and hardware end items are analyzed, and the resulting test case descriptions, test procedures, and test data are defined and captured in the plan. The software organization supports the development and review of the test plans and procedures. Dry runs of the system tests are performed to verify that the system test cases and test procedures are accurate and that the system is ready for formal, run-for-record testing.

Software problems encountered during the integration and testing are captured and resolved as described in the Software Configuration Management Plan (SCMP). If the analysis of the anomaly is determined to be a software issue, then the system test results are further reviewed and analyzed by the software configuration control board as described in the Software Configuration Management Plan. Once the software developers update and retest the software, the software baseline is updated and made available for retesting. Analysis of the tests and procedures helps determine which test procedures are candidates for retesting or regression testing. Regression testing ensures that the changes in the software/hardware correct the anomaly and do not degrade the system, and that the system still performs its overall requirements. Test engineers may also identify additional tests due to the changes.

B.3.12 Software Maintenance

This section describes the approach to be followed for maintaining the software end item(s) after completion of software formal testing. Once the software has completed software formal testing, software maintenance is performed in accordance with -07-10010, Maintain Developed Software method. The software end items and software work products continue to be updated and delivered as changes to the software are approved and implemented. The Software Review Plans are performed in order to determine the quality of the software work product before it is re-released.

Maintenance activities for the software end items are based on block updates to the software using a waterfall lifecycle model. Prior to the block update cycle, an analysis is conducted to determine the software problem reports to be incorporated into the block update. The initial software problem report selection for the software release update is conducted via the software change control board. Final approval of the changes to be incorporated into the release update is coordinated. The approved release of software problem reports implemented impacts software work product deliveries. In order to determine the quality of software changes before the release of software for system test and delivery to the program or project sponsor will require approval.

B.3.13 Software Delivery and Transition

This section describes the approach to be followed for preparing for software use and transitioning it to the project sponsor/stakeholder or its representative. This section includes the approach for delivering the software and performing the software installation and training at the user sites as specified in the contract.

For preparing the executable software, support files, and the corresponding VDD are described in the SCMP. The software development organization prepares and submits a software build request to SCM. After the software is built, the software development team executes the software prior to delivery to ensure that the executable code loads and runs. Then, the executable file(s) and any support files necessary to execute the software are delivered according to the baseline schedule. Software Configuration Management (SCM) provides the executable software, support files, and VDD for all deliveries. The software user documentation is developed to support the end users who will be operating the software and interpreting the results and data produced by the software. Once the software has been delivered, the software development team supports the installation and check-out of the software as well as user training. The transition plan describes the approach for transferring to the operational system and can be found in the Transition Plan. The software development team provides input to the development of the transition plan, supports the review of the plan, and participates in the transition activities as described in the plan.

B.4 SUPPORTING SOFTWARE PLAN

B.4.1 Software Configuration Management Plan

This section describes the approach to be followed for software configuration management, the Software Configuration Management Plan. The contents of this section are developed in accordance with the SCMP.

Software Configuration Management is performed in accordance with the Software Configuration Management Plan. The SCMP implements the configuration management policies for software. The SCMP describes the use of software configuration management procedures. The software manager, the software configuration management manager, and the team leader of the software and software configuration management organizations approve the SCMP.

B.4.2 Software Quality Assurance Plan

This section describes the approach to be followed for software quality engineering. Software Quality Engineering is performed in accordance with the Software Quality Assurance Plan (SQAP). The SQAP implements the quality policies for software and describes the use of software quality engineering procedures. The software manager, the software quality engineering manager, and the team leader of the software and software quality engineering organizations approve the SQAP.

B.4.3 Software Product Peer Reviews

This section is divided into subsections that describe the approach to be followed for software product reviews. The participants involved in the software product peer review are described in this section in order to ensure that the software work product is evaluated independently of the developer of the software work product.

Software work product peer reviews are conducted during the software lifecycle phase in which the product is developed. Checklists are provided for each software work product. These checklists are used during the review of the software work product in order to aid in the evaluation of the quality of the software work product. Once the peer review is completed, including the resolution of any findings, the software work product is submitted for a software lifecycle phase review or is placed under the level of configuration control. The process for release of a software work product, including the final approval of the product, is described in the Software Configuration Management Plan. Reviews of these products are performed in other forums, such as the change control board, management status reviews, and software quality product audits.

B.4.4 Software Technical Reviews

This section describes the software organization's approach for planning and participating in joint technical reviews at locations and dates proposed by the developer and approved by the acquirer. These reviews are commonly known as Technical Interchange Meetings (TIMs). Persons with technical knowledge of the software products to be reviewed attend these reviews. The reviews focus on in-process and final software products, as opposed to materials generated especially for the review.

TIMs are conducted periodically to do the following:

- Review evolving software products.
- Demonstrate that relationships, interactions, interdependencies, and interfaces between required software end items and system elements have been addressed.
- Resolve conflicts or inconsistencies among system requirements, functional alternatives, and design solutions.
- Assess and resolve technical issues associated with the software system and architecture.
- Surface near-term and long-term risks and arrive at agreed-upon strategies to minimize the identified risks.
- Ensure ongoing communications between the project sponsor/stakeholders and the developer technical personnel.

Preparation for the reviews includes schedule, agenda, and scope coordination with the appropriate attendees. Reviews may be conducted both electronically and as face-to-face meetings. Technical staff and managers review material internally before it is released to external participants. If necessary, a dry run is conducted to facilitate coordination, to ensure review readiness, and to ensure compliance with requirements. Review minutes, decisions, agreements, and approved action items are recorded and signed by the appropriate managers. Any resulting action items are reviewed periodically and tracked to closure.

B.4.5 Software Management Reviews

This section describes the approach for the software organization to plan and take part in software management reviews at locations and dates proposed by the developer and approved by the project sponsor/stakeholders. Persons with the authority to make cost and schedule decisions attend these reviews and have the following objectives:

- Keep management informed about project status, directions being taken, technical agreements reached, and overall status of evolving software products.
- Arrive at agreed-upon mitigation strategies for near- and long-term risks that could not be resolved at joint technical reviews.
- Identify and resolve management-level issues and risks not raised at technical reviews.
- Obtain commitments and project sponsor/stakeholder approvals needed for timely accomplishment of the project.

B.4.5.1 Software Lifecycle Phase Reviews

This section describes the approach for the software organization to plan and conduct the software lifecycle reviews that occur either at the beginning or at the conclusion of a software lifecycle phase. The locations and dates proposed by the developer and approved by the customer are determined for the reviews. Persons with the authority to make technical and authority decisions in order to provide approval for beginning the next software lifecycle phase attend these reviews. These reviews may be combined or conducted in conjunction with the system lifecycle reviews. The details for the reviews, including the entrance and exit criteria, are described.

Software lifecycle phase reviews are specified in Table B.1, Software Lifecycle Phase Reviews.

B.4.6 Software Supplier Management Plan

This section describes the approach to be followed for software supplier source selection and technical monitoring. The software organization

TABLE B.1

Software Lifecycle Phase Reviews

Review	Software Products Reviewed	Purpose
Software Specification Review	SRS, SDP, SCMP, SSP, SDEP, SQAP	Conducted for a software end item, or functionally related group of software end items, to evaluate the definition of the software requirements that have been allocated and derived from system requirements. System definitions and interface requirement specifications should be included. The review determines whether performance requirements for software are sufficiently defined for preliminary design to proceed.
Preliminary Design Review (PDR)	Preliminary SDD, Preliminary IDD, STP	Conducted for a software end item, or functionally related group of software end items, to evaluate the preliminary design in the SDD and preliminary IDD against the requirements allocated from the higher-level documents (i.e., SRS and IRS) to demonstrate that the software requirements have been satisfied. The software architecture is reviewed and evaluated to determine whether the approved design methodology has been implemented correctly. During the PDR, a technical understanding is reached on the validity and the degree of completeness of the software architecture. The test plans are reviewed to demonstrate that adequate test criteria for each software end item have been established and address all specified requirements. A PDR is conducted at the conclusion of the preliminary design activities prior to the start of detailed design.

Continued

TABLE B.1 (*Continued*)

Software Lifecycle Phase Reviews

Review	Software Products Reviewed	Purpose
Critical Design Review (CDR)	Final SDD, Final IDD, Preliminary STD	Conducted for a software end item, or functionally related group of software end items, to evaluate the detailed design documented in the updated SDD and updated IDD against the requirements allocated from the higher-level documents (i.e., SRS and IRS) to demonstrate that all software requirements have been satisfied. The low-level structure of the software end item is reviewed and evaluated to determine whether the approved design methodology has been implemented correctly and whether the design is complete. During the CDR, a determination is reached on the consistency of the updated SDD and the acceptability of the test cases for the formal qualification tests. A CDR is conducted at the conclusion of the detailed design activities prior to the start of coding and testing.
Test Readiness Review (TRR)	STP, Final STD, VDD (SDF showing all unit, integration tests completed)	Conducted for each software end item, or functionally related group of software end items, to determine whether the software formal testing procedures are complete and accurate and that the software is ready for formal testing. The software test descriptions are evaluated for compliance with the software testing plans and for adequacy in accomplishing testing requirements. A TRR is conducted after software testing procedures are available and software integration testing is complete. A successful TRR is predicated on a determination that the test procedures and results of software unit and integration testing form a satisfactory basis for proceeding into formal software end item testing.

| System-level final audit reviews identified in the plans such as Functional Configuration Audit (FCA) and Physical Configuration Audit (PCA) | Work products to be provided for the reviews such as SRS, SDD, STD, STR, VDD | FCA: This audit is conducted by the project sponsor/stakeholders to verify that the actual performance of the software end item complies with the requirements specified in the SRS. This is accomplished through a detailed review of the software formal test procedures, data, and reports. In addition, a technical understanding is reached on the validity and the degree of completeness of the operational and support documents. An FCA may be conducted at the completion of the formal software end item testing or after system integration and testing. PCA: This audit is conducted by the project sponsor/stakeholders to verify that the as-built version of a software end item reflects an up-to-date physical representation of the specifications and approved design of the software end item. The PCA includes a detailed audit of engineering drawings, specifications, technical data, and tests utilized in production of the hardware media for the software end item. The PCA also includes a detailed audit of the software end item requirements, design, and its configuration specified in the VDD. A PCA may be conducted at the completion of the formal software end item testing in conjunction with the FCA or after system integration and testing. The software end item product baseline is established upon the satisfactory completion of the FCA and PCA. |

supports supplier management during this process of defining the software supplier statement of work inputs and of evaluating and selecting candidate software suppliers. The work products documenting the software supplier source selection activities are developed and maintained by the supplier management organization.

B.4.7 Notes

This section contains any general information that aids in understanding this document (e.g., background information, glossary, rationale). This section includes a reference to the alphabetical listing of all acronyms, abbreviations, and their meanings as used in this document, and a list of any terms and definitions needed to understand this document.

Glossary

CDRL	Contract Data Requirements List
CM	Configuration Management
CMMI	Capability Maturity Model Integrated
COTS	Commercial-Off-the-Shelf
FCA	Functional Configuration Audit
IDD	Interface Design Document
PCA	Physical Configuration Audit
SCMP	Software Configuration Management Plan
SDE	Software Development Environment
SDF	Software Development File
SDP	Software Development Plan
SE	Systems Engineering
SEI	Software Engineering Institute
SEMP	Systems Engineering Management Plan
SOW	Statement of Work
SQA	Software Quality Assurance
SQAP	Software Quality Assurance Plan
SRD	Software Requirements Definition
SSR	Software Specification Review
STD	Software Test Description
STP	Software Test Plan
STR	Software Test Report
VDD	Software Version Description
TRR	Test Readiness Review

Terms

Software Configuration Management Plan
Software Quality Assurance Plan

The following terms are used throughout the document for the common organizational processes:

Software Requirements Development
Software Design
Software Coding
Software Unit Test
Software Integration and Integration Testing
Software Formal Test
Software User Documentation
Maintain Developed Software
Software Supplier Source
Software Supplier Technical Management
Software Architecture Development
Build Request
Software Change Request
Version Description Document
Software Metrics
Software Development Plans
Software Development Files
Software Requirements
Software Architecture Definition
Software Quality Plans
Software Requirements Verification
Software Defects
Software Quality Audit Performance
Perform Peer Reviews

Appendix

Appendices provide additional information that aids in understanding this document (e.g., notes, background information, rationale, definitions, etc.). Appendices will be referenced in the main body of the document.

Active Page Record

Page Numbers	Revision Level	Revision Type (Added, Deleted)	Page Numbers	Revision Level	Revision Type (Added, Deleted)

Revision Record

Revision Letter

Changes in This Revision

Authorization for Release

AUTHOR: _____   _____   _____
                  Signature   Org.   Date

APPROVAL: _____   _____   _____
                  Signature   Org.   Date

DOCUMENT RELEASE: _____   _____   _____
                  Signature   Org.   Date

Appendix C: Software Quality Assurance Plan

DOCUMENT NUMBER: **REVISION:**   **REVISION DATE:**

SQAP Information

Original Release Date	

Signatures for SQAP Approval and Release

AUTHOR: _____ _____ _____
Signature   Org.   Date

APPROVAL: _____ _____ _____
Signature   Org.   Date

DOCUMENT
RELEASE: _____ _____ _____
Signature   Org.   Date

Contents

List of Figures

List of Tables

Abstract

This document describes the Software Quality Program Plan and approach for providing Software Quality Assurance activities and tasks.

KEY WORDS

Commercial-Off-the-Shelf
Functional Configuration Audit
Physical Configuration Audit
Software Quality Assurance
Software Quality Program Plan
Software Development Plan

C.1 INTRODUCTION

C.1.1 Identification

This section is the Software Quality Assurance Plan (SQAP) System Overview.

C.1.2 Document Overview

This section describes how the SQAP is prepared and establishes the Software Quality Assurance (SQA) processes to be implemented into the following sections:

- Section 1 contains the overview information.
- Section 2 contains the documents referenced within this plan.
- Section 3 contains the SQA organizational structure and resources required for use on the software quality program.
- Section 4 contains or references the SQA procedures and tools employed.

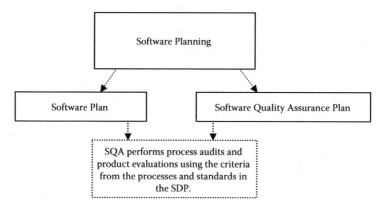

FIGURE C.1
Relation to other planning.

C.1.3 Relationship to Other Plans

This section explains that SQAP has a complementary relationship with the Software Development Plan (SDP). The SDP describes the procedures for the software development effort. The SDP is reviewed for compliance to contract requirements and standard software processes, and is approved by the SQA. This SQAP implements software quality to monitor the software development as identified in the SDP. Refer to Figure C.1, Relation to other planning.

C.2 REFERENCED DOCUMENTS

C.2.1 Documents

This section describes the nonapplicable documents that form a part of this document to the extent specified herein. In the event of conflict between the documents referenced herein and the contents of this document, this document will be considered a superseding requirement. Unless specified, the issues of these documents are those listed in the Index of Specifications and Standards. The documents are listed in Table C.1 for guidance and identify the documents referenced in this SQAP.

Copies of specifications, standards drawings, and publications required by suppliers in connection with specified procurement functions should be obtained from the contracting agency or as directed by the contracting officer.

TABLE C.1

Documents

Document Title	Version	Date
Software Quality Program Plan Data Item Description (DID)		mm/dd/yy
Software Systems and Equipment		mm/dd/yy

TABLE C.2

Nonapplicable Documents

Document Title	Version	Date
Software Quality Assurance Site Plan	Latest	mm/dd/yy

C.2.2 Nonapplicable Documents

This section describes the nonapplicable documents that form a part of this document to the extent specified herein. In the event of conflict between the documents referenced herein and the contents of this document, this document will be considered a superseding requirement. Table C.2 identifies the nonapplicable standards referenced in this document.

Copies of specifications, standards drawings, and publications required by suppliers in connection with specified procurement functions should be obtained from the contracting agency.

C.3 ORGANIZATION AND RESOURCES

C.3.1 Organization

This section states that SQA is the group responsible for implementing and ensuring compliance with Software Quality Assurance program requirements. Although SQA is a participating member of the Integrated Product Teams, all SQA personnel will have the resources, responsibility, authority and organizational freedom to permit objective process audits and product evaluations and to initiate and verify correction and corrective actions. The individuals performing software quality activities are not those who developed the product or are responsible for the product or activity. As shown in Figure C.2, SQA

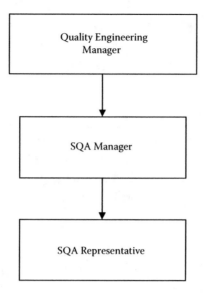

FIGURE C.2
SQA organizational structure.

maintains its independence by reporting directly to the SQA Manager, who has a reporting line to the Quality Engineering Manager. Since the purpose of this figure is to reflect the organizational structure, it will be updated for name changes only.

C.3.2 Resources

This section provides resources to fulfill the objectives of this SQAP. Program management has provided SQA funding based on software engineering hours.

C.3.2.1 Facilities and Equipment

This section describes SQA activities and how they apply to work being performed at the facility on site. The site facility houses all of the office and laboratory space required to successfully implement a software quality program. SQA is collocated within the program area to ensure cognizance of program activities.

Describe any unique facilities, including software library and resources. If SW is to be supplier developed, state this along with any facilities at the program site that will be utilized.

C.3.2.2 Facilities, Equipment, Software, and Services

This section describes the requirements for furnished facilities, equipment, software, or services to be used by SQA.

C.3.2.3 SQA Personnel

This section explains the number of Software Quality Assurance engineer(s) assigned. It is the responsibility of these individuals to implement and maintain the software quality program in accordance with this SQAP. If needed, additional support and training will be available from the SQA group.

C.3.2.4 Other Resources

This section explains how the Software Quality engineer is provided with various applications and electronic mail. Section C.4.2.2 identifies the Commercial-Off-the-Shelf (COTS) software that SQA will use in support of software quality tasks.

C.3.3 Schedule and Lifecycle Milestones

This section defines the SQA schedule and identifies the process audits to be performed by SQA on this program, which are based upon events identified in the SDP and the project schedule. This working master evaluation schedule is maintained by SQA.

The program's defined lifecycle identifies the milestones used by SQA in generating the SQA evaluation schedule. These milestones establish the beginning of specific activities and processes and their associated products:

- System Specification Review (SSR)
- Preliminary Design Review (PDR)
- Critical Design Review (CDR)
- Test Readiness Review (TRR)
- Formal Qualification Test (FQT)
- Functional Configuration Audit (FCA)/Physical Configuration Audit (PCA)

The major SQA milestones are as follows:

- Start of Software Planning
- Completion of Software Requirements Definition
- Completion of Software Design Definition
- Completion of Code and Unit Test
- Completion of Informal Integration Testing
- Start of Formal Qualification Test
- Software Delivery

C.4 SQA PROCEDURES, TOOLS, AND RECORDS

C.4.1 Procedures

This section explains the processes in the SQA organizational process set that describe the practices necessary for establishing and implementing a software quality program.

The following sections will identify the SQA Organizational Process Set to be implemented. Work Instructions for project-level tailored processes will identify the procedures for implementing SQA processes. SQA management approves project-level tailoring and reviews it for consideration.

C.4.1.1 Software Quality Program Planning

This section describes SQA activities used to plan, budget, and generate an SQAP.

The planning begins as early as possible in the proposal phase of a software development program. Ongoing coordination between the proposal team, the software group and the overall quality organization for new and ongoing activities occur with SQA. SQA planning and budgeting is an integral part of all new software development programs from the onset, and late involvement of SQA should be avoided. Qualified SQA personnel or SQA expertise is used to plan and ensure that adequate funding and resources are in place to properly sustain the SQA program plan.

Proposals include first-time proposal programs and updates to the budget cycle of ongoing programs.

C.4.1.2 Software Process Audit

This section describes the SQA activities used to conduct audits of software processes. SQA conducts these audits at planned intervals during the software lifecycle to verify conformance to tailored procedures.

C.4.1.3 Software Product Evaluation

This section describes the SQA activities used to objectively evaluate software products. SQA product evaluations may be performed and documented as a separate activity, or performed and documented as part of the Engineering Peer Review using the Peer Review process and tools.

C.4.1.4 SQA Support of Software Qualification

This section describes the SQA activities performed in support of software qualification. As detailed in Table C.3, SQA will ensure that software qualification testing is independently monitored and witnessed to verify that the test environment is configured, recorded, and controlled during testing, the testing is performed according to the test procedures, and all test records are accurate and controlled.

C.4.1.5 Support Software Configuration Management

This section describes the SQA activities used to support the Software Configuration Management process. These activities are performed in addition to the normal process and product audits and include participation in software review/change control boards, software master verification, and configuration verification for delivery. To satisfy configuration verification for delivery, SQA verifies completion of configuration verification audit, or equivalent, prior to delivery of a contractual software end

TABLE C.3

Plan to Support Software Test

Software Test	Witnessed or Monitored	Which Function (SQA or Other)?
Insert each planned formal software test, or just say ALL if appropriate	Indicate if test will be witnessed or monitored, or other	Indicate which function will witness or monitor (SQA or other)

TABLE C.4

Software Formal Reviews

Formal Review	Entrance Criteria	Exit Criteria
Identify planned software formal reviews for the program.	Identify entry criteria for the review.	Identify exit criteria for the review.

item. The audit provides evidence of as-designed/as-built compliance of the deliverable item.

C.4.1.6 Support Milestone Reviews

This section explains the SQA verification entry and exit criteria for milestone reviews based on contractual requirements and program standards. SQA participates in/supports the reviews identified in Table C.4.

C.4.1.7 Monitor Supplier Software Quality

This plan describes the SQA activities used to support software supplier management. This is applicable to suppliers delivering software that is developed or modified or delivered separately or embedded as part of a system (firmware on hardware).

The specific plan for monitoring software suppliers is documented in Table C.5.

C.4.1.8 Noncompliance Handling

This section explains how SQA initiates a closed-loop corrective action system when SQA or other members of the project identify noncompliance issues. Immediate correction addresses the removal of observed noncompliance, whereas long-term corrective action addresses the root-cause activities to prevent recurrence. Preventive actions are performed to prevent a potential noncompliance from occurring.

TABLE C.5

Plan for Monitoring Software Suppliers

Supplier	Project-Specific Audits Planned?	SQA Attendance at Milestone Reviews?	SQA Witness Formal Software Testing?
Insert names	Yes or no		
Bi-annual	Yes or no	Yes or no	

Noncompliance issues can be identified from various activities, including engineering peer reviews, testing, and SQA process audits or final product inspections. As a result, corrective action will sometimes be addressed by an engineering problem reporting and corrective action system. This section will only address the process for SQA noncompliance handling.

Implementation of this process can be scaled commensurate with the significance and severity of the finding. Waivers for noncompliance issues may only be approved by an established process, or engineering and quality management, and must be documented.

C.4.1.9 Reporting

This section explains how SQA reports software quality program activities, status, and results. It is important for SQA to provide the status of quality to management levels that have the responsibility to assess risk and make changes to the software program.

C.4.1.10 Records

This section defines how SQA maintains complete records of defined SQA activities in sufficient detail such that status and results are known as the following:

- List SQA records (e.g., checklists, audit reports, audit findings, and activity reports).
- Retention of records identified as Quality Records will follow the Master Records Retention Schedule.

C.4.2 Tools

This section defines the use of tools and provides a consistent, systematic approach to SQA principles. SQA utilizes the tools described in the following subsections.

C.4.2.1 Checklists

This section describes the standard checklists obtained from the SQA repository and is updated as necessary to reflect project tailoring. Although

TABLE C.6

COTS Software Usage

Product	Version	Use
Microsoft Outlook	2003	To provide intergroup communication
Microsoft Excel	2003	To generate databases and spreadsheets
Microsoft Access	2003	To maintain organizational data
Internet Explorer	7	For all Web-based applications and data
Microsoft Word	2003	Word processing
Microsoft PowerPoint	2003	To generate SQA presentation slides

checklists may guide an evaluation, they do not limit it. Checklists are attached to the SQA audit/evaluation report and are maintained as part of the SQA records.

C.4.2.2 COTS Software

This section explains how COTS software, as identified in Table C.6, will be used by SQA in support of tasks identified in this plan. The product version and product use is also identified. Note that this table will not reflect any upgrades to newer versions of these items.

Glossary

CDR	Critical Design Review
COTS	Commercial-Off-the-Shelf (Software)
PDR	Preliminary Design Review
FCA	Functional Configuration Audit
FQT	Formal Qualification Test
PCA	Physical Configuration Audit
SDP	Software Development Plan
SQAP	Software Quality Program Plan
SQA	Software Quality Assurance
SSR	System Specification Review
SW	Software
TRR	Test Readiness Review

Terms

The following terms are used throughout the SQAP.

Engineering Review Board
Software Integration and Integration Testing
Software Formal Test
Software User Documentation
Maintain Developed Software
Software Supplier Source
Software Supplier Technical Management
Software Architecture Development
Build Request
Software Change Request
Version Description Document
Software Development Plans
Software Requirements
Software Architecture Definition
Software Quality Plans
Software Requirements Verification
Software Defects
Perform Peer Reviews

Active Page Record

Active Page Record

Page Numbers	Revision Level	Revision Type (Added, Deleted)	Page Numbers	Revision Level	Revision Type (Added, Deleted)

Revision Record

Revision Letter

Changes in This Revision

Authorization for Release

AUTHOR: _____ _____ _____
 Signature Org. Date

APPROVAL: _____ _____ _____
 Signature Org. Date

DOCUMENT
RELEASE: _____ _____ _____
 Signature Org. Date

Appendix D: Software Configuration Management Plan

DOCUMENT NUMBER: **REVISION:**   **REVISION DATE:**

SCMP Information

Release Date	

Signatures for SDP Approval and Release

AUTHOR: _____ _____ _____
                    Signature   Org.   Date

APPROVAL: _____ _____ _____
                    Signature   Org.   Date

DOCUMENT
RELEASE: _____ _____ _____
                    Signature   Org.   Date

Contents

List of Figures

None

List of Tables

None

Abstract

This document describes the Software Configuration Management Plan and approach for providing Software Configuration Management activities and tasks.

KEY WORDS

Commercial-Off-the-Shelf
Computer Program Library
Engineering Review Board
Software Change Request
Software Configuration Management Plan
Software Quality Program Plan
Software Development Plan

D.1 INTRODUCTION

This section provides a high-level overview describing the purpose and structure of the SCMP. It identifies the software end items and work products to which the plan is applicable, and it defines the relationship of the SCMP to other plans (program plans, software plans, etc.).

D.1.1 Purpose

Provide a high-level description of the SCMP.

D.1.2 Scope

Identify the applicability of the SCMP.

D.1.3 Software Products

Identify the software to which this SCMP will apply.

D.1.4 Document Overview

Provide a high-level description of how the SCMP is defined.

D.1.5 SCMP Maintenance

Describe the conditions under which this plan is updated and how changes to it are approved and implemented.

D.1.6 Relationship to Other Plans

Describe the relationship of the SCMP to other plans.

D.2 ORGANIZATION AND RESOURCES

This section is divided into subsections that provide a description of the organization's structure and SCM responsibilities:

- Resources, tools, and training to accomplish SCM tasks.
- Tasks are estimated, budgeted, planned, and measured.
- Major milestones and tasks associated with the SCM processes are described in this SCMP.

D.2.1 Other Organizations

Identify the organization's structure and SCM responsibilities and its relationship to other organizations.

D.2.1.1 SCM Organization

Identify the SCM organization and its responsibilities.

D.2.1.2 SCM Relationships

Describe the relationship between the SCM organization and other organizations.

D.2.2 SCM Resources

Identify the resources used to perform the SCM tasks defined in the SCMP.

D.2.2.1 SCM Personnel

Identify SCM personnel staffing.

D.2.2.2 SCM Tools

Identify the SCM tools to perform the SCM tasks defined in the SCMP.

D.2.3 SCM Management

Describe how SCM is budgeted and scheduled and how SCM performance is measured and status is reported.

D.2.3.1 SCM Estimation, Budget, and Planning

Describe the SCM responsibilities and provide estimation, budget, and project plans.

D.2.3.2 SCM Reporting

Describe how the SCM group tasks are reported.

D.2.4 SCM Training

Describe the SCM training that is provided to SCM for performing their tasks.

D.2.5 SCM Process Reviews

Describe the SCM process reviews that are conducted to ensure compliance with the processes described in the SCMP.

D.2.6 SCM Milestones and Activities

Outline the major SCM milestones and activities provided as described in the SCMP.

D.3 CONFIGURATION IDENTIFICATION

This section is divided into subsections that describe configuration identifications per the following:

- Data and documentation that will be used to define the configuration
- Baselines established
- Identification of software products
- CPL processes and procedures
- SCM builds
- Controlled software releases

D.3.1 Data and Documentation

Describe the data and documentation used to define and maintain the configuration control. This includes the following:

- Requirements, design, code and unit test, as-built configurations, installations, maintenance information
- Enforcement of software standard(s) that will be used that define the requirements utilized for data/documentation control

D.3.2 Software Baselines and Documentation

Describe the software baselines and documentation to implement configuration control.

D.3.3 Identification Methods

Describe the identification methods that are used to uniquely identify the software and its associated work.

D.3.3.1 Software Product Identification

Describe the method used to provide product identification.

D.3.3.2 Configuration Index

Document the configuration index to identify software products.

D.3.3.3 Version Description Document

Identify the Version Description Document (VDD) that is used to document the configuration of software products formally delivered to sites, labs, and the customer.

D.3.4 Computer Program Library

Describe the program and IPT software library structure and how it is used for development, control, retention, distribution, and delivery of software products.

Describe access and physical controls utilized to ensure the integrity of library contents.

D.3.4.1 Distribution and Delivery

Describe distribution between the CPL and the delivery of the software products.

D.3.4.1.1 Computer Program Library Distribution

Describe how software products are checked out from the CPL to support software development activities.

D.3.5 Software Builds

Describe how software builds (engineering, informal, and formal) are performed.

D.4 CONFIGURATION CONTROL

Describe how configuration control is implemented:

- Type of configuration controls
- Change documentation and accountability
- ERB responsibilities
- Software configuration control
- Configured software releases

D.4.1 Configuration Control Levels

Identify the levels of control that are applied to the software products.

D.4.1.1 Change Type and Classification

Identify and describe the types of changes associated with each level of control.

D.4.2 Software Change Request

Provide a description of the Software Change Request (SCR) and how it is used to identify, control, and track software changes.

D.4.3 Engineering Review Board

Describe the ERB, its membership, and authority. Describe the relationship between the ERB and higher-level Configuration Control Boards (CCBs).

D.4.4 Software Product Release

Describe how the release of software products is controlled.

D.4.5 Additional Controls

Identify controls needed for software items.

D.5 CONFIGURATION STATUS ACCOUNTING

Configuration status accounting records and status to support configuration audit activities.

- SCM records (e.g., CPL records, audit records, ERB minutes, and agendas)
- SCM status accounting records and reports (change status reports, audit records, reports)
- Metrics created and prepared

D.5.1 SCM Records

Identify SCM records used to support verification, validation, and audits.

D.5.2 SCM Reports

Identify SCM reports used to support verification, validation, and audits.

D.5.3 SCM Prepared Metrics

Identify SCM software metrics and reports.

D.6 CONFIGURATION AUDITS

This section is divided into subsections that describe how configuration audits are implemented on the program to ensure that

- Software data and documentation reflect the as-built configuration.
- Approved changes are implemented into the software products.
- Products comply with applicable standards.

- SCM process compliance.
- Computer Program Library integrity.

D.6.1 SCM Internal Audits

Describe the internal configuration audits performed by SCM to ensure configuration control.

D.6.2 Configuration Audits

Describe the configuration audits that will be performed with the customer to verify the configuration and performance of a software product.

D.7 INTERFACE CONTROL

Describe how interfaces between software products and other components of the system are identified and controlled.

D.8 SUPPLIER MANAGEMENT

This section describes the applicability of SCM to suppliers/subcontractors. Include unique delivery requirements (electronic, paperwork, etc.).

Glossary

CCB	Configuration Control Board
CDR	Critical Design Review
COTS	Commercial-Off-the-Shelf (Software)
CPL	Computer Program Library
ERB	Engineering Review Board
SCM	Software Configuration Management
SCMP	Software Configuration Management Plan
SCR	Software Change Request
SDP	Software Development Plan
SQAP	Software Quality Program Plan
SQA	Software Quality Assurance
SW	Software
TRR	Test Readiness Review
VDD	Version Description Document

Terms

The following terms are used throughout the SCMP.

Computer Program Library
Configuration Control Board
Engineering Review Board
Software Formal Test
Maintain Developed Software
Software Supplier Source
Build Request
Software Change Request
Version Description Document
Software Development Plans
Software Requirements
Software Quality Plans

Active Page Record

Page Numbers	Revision Level	Revision Type (Added, Deleted)	Page Numbers	Revision Level	Revision Type (Added, Deleted)

Revision Record

Revision Letter

Changes in This Revision

Authorization for Release

AUTHOR: _____ _____ _____
                    Signature   Org.   Date

APPROVAL: _____ _____ _____
                    Signature   Org.   Date

DOCUMENT RELEASE: _____ _____ _____
                    Signature   Org.   Date

Appendix E: FCA Checklist

CSCI and Number	Date Started:
Procurement Package	Date Completed:

Documentation Baseline (SDRLs) Engineering Data Package

Document	SDRL	Document Number/Rev	Yes	No
ICD	001			
SCMP	002			
SQAP	003			
Supplier Change Proposal	004			
SDP	005			
VDD	006			
SUM	007			
SDD	009			
SRS	010			
IRS	011			
Software Product End Items	012			
STP	013			
STD	014			
IDD	015			
SCR Data	016			
SIP	017			
SCMP	018			
SDD	019			
Correct Actions	020			
Drawing Package	021			
Document Package	022			
COTS Items	022			

Delivery from SOW

Software Media:

Software Item	Comments

Activities	Comments
Entry Criteria: • Software Readiness Review held • FCA Audit Team set up and notified of audit • FCA agreed-to agenda • FCA responsibilities defined for the Audit Team • FCA Checklist completed • FCA minutes available	
Product Configuration Baseline identified: • Operating and software support documents were reviewed (SUM, Firmware Support Manual, Operating Manuals) • The (Software) Subcontract Management Plan is released • Test software identified in VDD with reference to media • List of approved software changes	
Specification review and validation to define the configuration item, testing, mobility/transportability, and packaging requirements: • Packaging plan/requirements review complete	
Drawing/document review of outstanding design changes, part numbers, and description: • Changes incorporated between the Test baseline and Release baseline • SW installation drawings/documents and inspections	

Activities	Comments
Review of unincorporated design changes: • Outstanding high-level and middle-level changes, SCRs	
Review waivers and deviations to specifications and standards: • Action Item database open items impacting software	
Review Engineering Release and Change Control System for control of processing and formal release of engineering changes: • Test tools identified • Software tools identified • Software part numbers • Review of software build processes • Software media labeling requirements • Software media storage requirements	
Software CSCI reviewed: • Software component design descriptions • Software interface requirements • Status of reviews (SRR, PDR, CDR, TRR) • Findings/status of SQA reviews • Subcontractors developing software • Software embedded COTS • Peer Review Report	
MIL-STD-498 compliance FCA – Yes or No	
Exit Criteria: • FCA roster • FCA agenda and presentation data • FCA Action Item Log and Action Items • Test data is complete and accurate • Verification tasks against requirements are complete	

FCA Action Items

AI Number	Description	ECD

Closing Comments:

The software performs as required by the allocated configuration to create the product baseline.

Yes _____

No _____

Appendix F: PCA Checklist

CSCI and Number	Date Started:
Procurement Package	Date Completed:

Documentation Baseline (SDRLs) Engineering Data Package

Document	SDRL	Document Number/Rev	Yes	No
ICD	001			
SCMP	002			
SQAP	003			
Supplier Change Proposal	004			
SDP	005			
VDD	006			
SUM	007			
SDD	009			
SRS	010			
STR	011			
Software Product End Items	012			
STP	013			
STD	014			
IDD	015			
SCR Data	016			
SIP	017			
SCMP	018			
SDD	019			
Correct Actions	020			
Drawing Package	021			
Document Package	022			
COTS Items	022			

Delivery from SOW

Software Media:

Software Item	Comments

Activities	Comments
PCA Entry Criteria met. • Readiness Review successfully held • Defined responsibilities and authority for PCA • PCA agreed-to agenda • Presentation including scope and in-brief materials • PCA Checklist completed • PCA presentation material clear and sufficient in detail and consistent within the scope of the PCA • PCA minutes available	
Product Configuration Baseline identified. • Were the operating and software support documents reviewed? (SUM, SPM, Firmware Support Manual, Operating Manuals) • The (Software) Subcontract Management Plan is released? • Test software identified in VDD with reference to media? • List of approved changes	

Activities	Comments
Specification review and validation to define the configuration item, testing, mobility/transportability, and packaging requirements. • Packaging plan/requirements review complete • Test procedures and results in compliance with software specifications requirements • Configuration items examined to ensure conformance with requirements • Software Product Specification reviewed pertaining to format and completeness • Interface Requirements reviewed and in compliance	
Drawing review against as-built and variations, including outstanding design changes, part numbers, and description. • Changes incorporated between Test baseline and Release baseline? • Installation drawings and inspections complete • Software user documentation checked for completeness and in conformance with applicable data items	
Review of unincorporated design changes. • Outstanding high-level and middle-level changes, SCRs	
Review waivers and deviations to specifications and standards. • Action Item database open items impacting software	

Activities	Comments
Review Engineering Release and Change Control System for control of processing and formal release of engineering changes. • Test Tools Identified • Software Tools Identified • SW part number • Review of build processes • Software media labeling requirements • Software media storage requirement	
Was the CSCI reviewed? • Software component design descriptions • Software interface requirements • Status of reviews (SRR, PDR, CDR, TRR) • Findings/status of SQA reviews • Subcontractors developing software • Embedded COTS • Peer Review Report	
Software Product Evaluations complete for each item subject to the PCA	
MIL-STD-498 compliance? PCA – Yes or No	
Exit Criteria • Roster • Agenda and presentation data • Action Item Log and Action Items • Data Package • Signed Certification Sheets • Test data is complete and accurate • Verification tasks against requirements are complete	

Certifications

Certification	Status
PCA Cover Sheet	
Certification Package Scope/Purpose	
Product Configuration Baseline established	
Specification Reviews and Validation	
Software Drawing/Document Review	
Unincorporated Software Design Changes	
Software Deviations/Waivers	
Engineering Release/Change Control	

Action Items

AI Number	Description	ECD

Comments:

The as-built configuration is established by released documentation to establish the product baseline.

Yes _____

No _____

Acronyms

A

AD: Airworthiness Directives
ANSI: American National Standards Institute
AS: Aerospace
ATP: Acceptance Test Procedure

C

CA: Corrective Action
CAP: Configuration Audit Plan
CASE: Computer Aided Software Engineering
CASR: Configuration Audit Summary Report
CSCI: Computer Software Configuration Item
CDR: Critical Design Review
CDRL: Contract Data Requirement List
CI: Configuration Item
CM: Configuration Management
CMM: Compatibility Maturity Model
CMMI: Compatibility Maturity Model Integration
COTS: Commercial Off the Shelf
CPR: Common Problem Report
CSA: Configuration Status Accounting
CSAS: Configuration Status Accounting System
CSC: Computer Software Component
CSU: Computer Software Unit

D

DBDD: Database Design Description
DBMS: Database Management System
DID: Data Item Description
DOD: Department of Defense
DOORS: Dynamic Object-Oriented Requirements System
DPR: Development Progress Review

E

EP: Execution Plans
EQM: Engineering Quality Manager
EML: Extensible Markup Language

F

FAI: First Article Inspection
FCA: Functional Configuration Audit
FDR: Final Design Review
FQT: Formal Qualification Test
HDP: Hardware Development Plan
HWCI: Hardware Configuration Item

I

ICP: Interface Control Plan
IDD: Interface Design Description
IDL: Interface Design Language
IEEE: Institute of Electrical and Electronic Engineers
IPMP: Integrated Performance Management Plan
IPT: Integrated Product Team

IRS: Interface Requirements Specification
IT: Information Technology
ITT: Integrated Test Team

K

KPA: Key Process Area

M

M&C: Managed and Controlled
MIL-STD: Military Standard
MLB: Major League Baseball
MRI: Master Record Index

N

NDI: Non-Developmental Item
NDS: Non-Developed Software
NFL: National Football League
NRO: National Reconnaissance Organization
NASA: National Aeronautics and Space Administration

O

OFI: Opportunity for Improvement

P

PCA: Physical Configuration Audit

PDR: Preliminary Design Review
PEP: Program Execution Plan
PM: Process Monitor
POL: Policy
PP&C: Program Planning and Control
PQP: Program Quality Plan
PR: Peer Review

Q

QA: Quality Assurance
QMS: Quality Management System

R

RSM: Risk Management System
RO: Release Order
RR: Release Record
RR: Readiness Review

S

SAC: System Acquisition Contract
SCM: Software Configuration Management
SCMP: Software Configuration Management Plan
SCR: System Change Request
SCSS: Specification Compliance Summary Sheet
SDD: Software Design Document
SDF: Software Development File
SDL: Software Developmental Lab
SDLIB: Software Development Library
SDP: Software Development Plan

SDR: System Design Review
SDRL : Supplier Data Requirements Lists
SEMP: System Engineering Management Plan
SIL: System Integration Lab
SI: Software Inspection
SMP: Subcontract Management Plan
SOW: Statement of Work
SEI: Software Engineering Institute
SPC: Statistical Process Control
SPO: System Program Office
SPR: Software Problem Report
SQA: Software Quality Assurance
SQE: Software Quality Engineering
SQAP: Software Quality Assurance Plan
SRN: Software Release Notice
SRR : System Requirements Review
SRS: Software Requirements Specification
SSP: Software Support Plan
SSR: Software Specification Review
STD: Software Test Description
STP: Software Test Plan
SW: Software

T

TPM: Technical Performance Measurement
TT&E: Tactical Test and Evaluation
TIM: Technical Interchange Meeting
TRR: Test Readiness Review

U

USAF: United States Air Force

V

VDD: Version Description Document

W

WBS: Work Breakdown Structure

Index